SACRAMENTO PUBLIC LIBRARY

3 3029 04822 2087

solutions@syngres

CENTRAL LIBRARY
828 "I" STREET
SACRAMENTO, CA 95814
FEB - - 2003

With more than 1,500,000 copies of our MCS study guides in print, we continue to look for information needs of our readers. One way we

Readers like yourself have been telling us they want an Internet-based service that would extend and enhance the value of our books. Based on reader feedback and our own strategic plan, we have created a Web site that we hope will exceed your expectations.

Solutions@syngress.com is an interactive treasure mation focusing on our book topics and related te offers the following features:

- One-year warranty against content obsolescence due to vendor product upgrades. You can access online updates for any affected chapters.

- "Ask the Author" customer query forms that enable you to post questions to our authors and editors.

- Exclusive monthly mailings in which our experts provide answers to reader queries and clear explanations of complex material.

- Regularly updated links to sites specially selected by our editors for readers desiring additional reliable information on key topics.

Best of all, the book you're now holding is your key to this amazing site. Just go to **www.syngress.com/solutions**, and keep this book handy when you register to verify your purchase.

Thank you for giving us the opportunity to serve your needs. And be sure to let us know if there's anything else we can do to help you get the maximum value from your investment. We're listening.

www.syngress.com/solutions

SYNGRESS®

YNGRESS®

1 YEAR UPGRADE
BUYER PROTECTION PLAN

10 COOL

LEGO®
Mindstorms™
ULTIMATE BUILDERS PROJECTS

Syngress Publishing, Inc., the author(s), and any person or firm involved in the writing, editing, or production (collectively "Makers") of this book ("the Work") do not guarantee or warrant the results to be obtained from the Work.

There is no guarantee of any kind, expressed or implied, regarding the Work or its contents. The Work is sold AS IS and WITHOUT WARRANTY. You may have other legal rights, which vary from state to state.

In no event will Makers be liable to you for damages, including any loss of profits, lost savings, or other incidental or consequential damages arising out from the Work or its contents. Because some states do not allow the exclusion or limitation of liability for consequential or incidental damages, the above limitation may not apply to you.

You should always use reasonable care, including backup and other appropriate precautions, when working with computers, networks, data, and files.

Syngress Media®, Syngress®, "Career Advancement Through Skill Enhancement®," and "Ask the Author UPDATE®," are registered trademarks of Syngress Publishing, Inc. "Mission Critical™," "Hack Proofing®," and "The Only Way to Stop a Hacker is to Think Like One™" are trademarks of Syngress Publishing, Inc. Brands and product names mentioned in this book are trademarks or service marks of their respective companies.

KEY	SERIAL NUMBER
001	66YH7V43GY
002	RT5N7TF6CV
003	QASHF35TXA
004	XC4H8976NY
005	H74PR33T5S
006	UJF7NCES64
007	EC34EQ2APK
008	A2DMRDKJ67
009	V8NKP7HV6F
010	W6XF79UM3Z

PUBLISHED BY
Syngress Publishing, Inc.
800 Hingham Street
Rockland, MA 02370

10 Cool LEGO® MINDSTORMS™ Ultimate Buider Projects

Copyright © 2002 by Syngress Publishing, Inc. All rights reserved. Printed in the United States of America. Except as permitted under the Copyright Act of 1976, no part of this publication may be reproduced or distributed in any form or by any means, or stored in a database or retrieval system, without the prior written permission of the publisher, with the exception that the program listings may be entered, stored, and executed in a computer system, but they may not be reproduced for publication.

Printed in the United States of America

1 2 3 4 5 6 7 8 9 0

ISBN: 1-931836-60-4

Technical Reviewers: Mario Ferrari and Giulio Ferrari   Cover Designer: Michael Kavish
Acquisitions Editors: Catherine B. Nolan and   Page Layout and Art by: Shannon Tozier
Jonathan E. Babcock   Copy Editor: Kate Glennon

Distributed by Publishers Group West in the United States and Jaguar Book Group in Canada.

Acknowledgments

We would like to acknowledge the following people for their kindness and support in making this book possible.

A special thanks to Matt Gerber at Brickswest for his help and support for our books.

Karen Cross, Lance Tilford, Meaghan Cunningham, Kim Wylie, Harry Kirchner, Kevin Votel, Kent Anderson, Frida Yara, Jon Mayes, John Mesjak, Peg O'Donnell, Sandra Patterson, Betty Redmond, Roy Remer, Ron Shapiro, Patricia Kelly, Andrea Tetrick, Jennifer Pascal, Doug Reil, David Dahl, Janis Carpenter, and Susan Fryer of Publishers Group West for sharing their incredible marketing experience and expertise.

Duncan Enright, AnnHelen Lindeholm, David Burton, Febea Marinetti, and Rosie Moss of Elsevier Science for making certain that our vision remains worldwide in scope.

David Buckland, Wendi Wong, Daniel Loh, Marie Chieng, Lucy Chong, Leslie Lim, Audrey Gan, and Joseph Chan of Transquest Publishers for the enthusiasm with which they receive our books.

Kwon Sung June at Acorn Publishing for his support.

Jackie Gross, Gayle Voycey, Alexia Penny, Anik Robitaille, Craig Siddall, Darlene Morrow, Iolanda Miller, Jane Mackay, and Marie Skelly at Jackie Gross & Associates for all their help and enthusiasm representing our product in Canada.

Lois Fraser, Connie McMenemy, Shannon Russell, and the rest of the great folks at Jaguar Book Group for their help with distribution of Syngress books in Canada.

David Scott, Annette Scott, Delta Sams, Geoff Ebbs, Hedley Partis, and Tricia Herbert of Woodslane for distributing our books throughout Australia, New Zealand, Papua New Guinea, Fiji Tonga, Solomon Islands, and the Cook Islands.

Winston Lim of Global Publishing for his help and support with distribution of Syngress books in the Philippines.

Contributors

Stephen Cavers began his secret life as a LEGO MINDSTORMS builder in March 2001, when he bought the Robotics Invention System 1.5. Since then, he has been slowly expanding his LEGO collection and his skills at building little plastic robots.

Before his MINDSTORMS addiction took hold, Steven received a bachelor's degree in Theatre at the University of British Columbia. His longtime interest in computers and technology led him into the software industry, where he has worked as a technical writer since 1994.

Stephen currently lives in Vancouver, Canada, and enjoys consuming copious amounts of sushi between LEGO projects.

Stephen is the creator of Robot 1: The WideBot, Robot 2: SumoBug, Robot 3: Hopper, and Robot 4: HunterBot.

Dr. Soh Chio Siong (commonly known as **CSSoh** on the Internet) is a Public Health Physician who has a penchant for things scientific, mechanical, and electronic. Since he was a child, he has built crystal sets, microscopes, telescopes, steam engines, digital clocks, and computers, among other things.

Dr. Soh became interested in using LEGO as a tool for creative learning in 1998, with the purchase of some LEGO Dacta sets and, later on, the MINDSTORMS RIS set. He developed a special interest in pneumatics, particularly pneumatic engines, and is author of the world-renowned site on LEGO Pneumatics (www.geocities.com/cssoh1). He is an active member of the LUGNET community and has led many interesting discussion threads.

His current interest is the use of LEGO in the teaching of science and creativity. He thinks robotics should be the fifth R, after Reading, wRiting, aRithmetic, and computeR. He lives with his wife and daughter in Singapore.

Other LEGO claims to fame for Dr. Soh include:

In September 1999, Dr. Soh's RCX Controlled Air Compressor Tester (www.lugnet.com/robotics/?n=7407) created quite a stir on the LUGNET Robotics Discussion list.

CSSoh's LEGO Pneumatics Page (www.geocities.com/cssoh1) was voted LUGNET's Cool LEGO Site of the Week for January 9–15, 2000. This was the first site from Singapore to receive this recognition from LUGNET.

In June 2000, Dr. Soh, in collaboration with P.A. Rikvold and S. J. Mitchell of Florida State University, participated in a poster presentation at the Gordon Conference. The presentation, entitled "Teaching Physics with LEGO: From

Steam Engines to Robots," can be viewed at www.physics.fsu.edu/users/rikvold/info/gordon00a.html.

Dr. Soh is the creator of Robot 5: Nessie, and Robot 6: Nellie.

David Astolfo recalls that Lego first stimulated his imagination at about 4 years old. It was not long before he received my first Lego Technic set. As a person who loves to take things apart, figure out how then work, and sometimes put them all back together, Technic was the ideal toy,. he would spend hours building various Technic trucks, cranes and tractors, only to tear them apart and start something new. One of his earliest Technic creations was a front wheel drive steering-capable mechanism built from the parts of the 853 Technic Auto Chassis set (1977).

A few years ago, David discovered that Lego had been working with MIT to produce a "smart brick"., he knew then, that his "dark period" was over and he was going to have to dust off the bricks and start building again. Soon after, he picked up his first MINDSTORMS Robotics Invention System set I could get my hands on. David now owns 5 RIS sets and a number of other Technic sets totaling a part inventory over 25000 pieces.

David lives with his wife Rebecca in Burlington Ontario Canada. Currently employed at ASI Technologies Inc. as Manager of Applications Systems, David's specialties include: Web application development and deployment, mapping with GIS software tools, database modeling & design and a variety of network infrastructure management tasks. His educational background consists of a Bachelor of Science Degree, as well as a Geographic Applications Specialist (GIS) Certificate.

When not working, David's hobbies consist of Karate, mountain biking, and creating robots with LEGO of course. He also occasionally attends robotic competitions that are held by the rtlToronto group in Toronto. This group offers some great challenges and a friendly and fun environment to test ones robot-building skills against others. For information on David's other Lego creations, visit his Web site at www.astolfo.com/bots.

David is the creator of Robot 7: The DominoBot.

Dr. Larry Whitman is an Assistant Professor of Industrial and Manufacturing Engineering at Wichita State University. He teaches and performs research in the areas of supply chain management, lean manufacturing, virtual reality, and computer integrated manufacturing. He uses LEGO to demonstrate production systems concepts to classes of college, high school, and middle school students. He has presented production system concepts using LEGO at industrial engineering and resource management national conferences. He is a den leader for his son, Joshua's Cub Scout pack and works with his son to build new designs and modify other LEGO designs. Larry spent ten years in the defense industry integrating factory automation and also integrating engineering computer aided design with manufacturing.

Dr. Whitman, in conjunction with Tonya and Alex Witherspoon, is the creator of Robot 8: The Drawbridge.

Tonya L. Witherspoon is an Educational Technology Instructor at Wichita State University (WSU) in Wichita, KS. She teaches clay animation, multimedia production, Web design, and several robotics and programming courses using the LEGO MINDSTORMS RIS, Logo, Handy Crickets, and Roamer robots. She has co-authored several books on integrating technology into curriculum, speaks at state and national conferences on the subject, and teaches workshops and in-service for many schools in Kansas.

Tonya's interest in robotics peaked during Mindfest, a forum hosted by the Massachusetts Institute of Technology (MIT) in October 1999. She was inspired when Dr. Seymour Papert spoke about his work with MINDSTORMS and challenged everyone to encourage learning and find ways to spread knowledge in new and exciting ways. Since then, Tonya has received two grants that allowed her to give teachers in Kansas a MINDSTORMS RIS kit upon completion of a robotics workshop at Wichita State University. To date, she's given away over 75 RIS kits and helped many teachers find funding for complete classroom sets. She hosted a robotics summer camp this past summer for over 65 middle-school students; the camp also served as a practicum for teachers to learn how to use the MINDSTORMS RIS in their classrooms. In collaboration with WSU's College of Engineering, she has hosted two annual MINDSTORMS Robotics Challenges, events in which over 200 middle-school students from Kansas have competed in robotic challenges. The third annual MINDSTORMS Robotics Challenge will be hosted in March 2003 (http://education.wichita.edu/mindstorms).

Tonya's family consists of her husband, Steve, who is a teacher, and five school-age children: Andrew, Alex, Adam, Austin, and Madeline. She resides in Wichita, but lives in cyberspace.

Tonya Witherspoon, in collaboration with her son, Alex, contributed Robot 9: The Wrapper Compactor. Tonya was also a special collaborator with Dr. Larry Whitman on Robot 8: The Drawbridge.

Alex Witherspoon is a middle-school student in Wichita, KS. His brain is hardwired for innovation; he has designed numerous creations on notebook paper since preschool. One of his first creations was a practical Midwestern solution: an explosive that would counteract and diffuse a tornado. Alex also designed a multilevel clubhouse, complete with a bed, television, computer, and a

McDonalds on the lower level. He has made that clubhouse a reality in his back-yard (minus the McDonalds). Alex presented his robot "Catapult Mania" at MIT's Mindfest when he was nine and broke the code to unlock the LEGO Knight's chain, which was a challenge posed to all Mindfest participants. His reward was to take home the four-foot LEGO Knight. Upon returning from Mindfest, Alex and his mother started a school-funded robotics club, at the invitation of Alex's elementary school principal.

The journey to MIT showed Alex that his type of creativity has ample appli-cation in our world, and has spawned different inventions using LEGOs and other materials to consummate the tenuous relationship between idea and reality. Alex has participated on robotics teams that have received the top prize for two years in a row at WSU's MINDSTORMS Robotics Challenge. This summer, he sent for a free patent attorney's kit.

Alex Witherspoon, in collaboration with his mother, Tonya, contributed Robot 9: The Wrapper Compactor. Alex was also a special collaborator with Dr. Larry Whitman on Robot 8: The Drawbridge.

Kevin Clague graduated in 1983 from Iowa State University with a bachelor's of Science degree in Computer Engineering. For the past 18 years, Kevin has worked as a Diagnostic Engineer at the Amdahl Corporation. For the last two years, he has also acted as a Senior Staff Engineer doing verification work at Sun Microsystems on their Ultra-Sparc V RISC processor. Kevin has two major hobbies: theatrical lighting and LEGO MINDSTORMS. Kevin has been playing with the RIS 1.5 for several years now and is currently working on LPub, an application to revolu-tionize the world of creating online LEGO building instructions.

Kevin Clague contributed Robot 10: Robo-Hominid.

Technical Reviewers

Mario Ferrari received his first LEGO box around 1964, when he was four-years-old. LEGO was his favorite toy for many years, until he thought he was too old to play with it. In 1998, the LEGO MINDSTORMS RIS set gave him reason to again have LEGO become his main addiction. Mario believes LEGO is the closest thing to the perfect toy. He is Managing Director at EDIS, a leader in finishing and packaging solutions and promotional packaging. The advent of the MINDSTORMS product line represented for him the perfect opportunity to combine his interest in IT and robotics with his passion for LEGO bricks. Mario has been an active member of the online MINDSTORMS community from the beginning and has pushed LEGO robotics to its limits. Mario holds a bachelor's degree in Business Administration from the University of Turin and has always nourished a strong interest for physics, mathematics, and computer science. He is fluent in many programming languages and his background includes positions as an IT Manager and as a Project Supervisor. With his brother Giulio Ferrari, Mario is the co-author of the highly successful book *Building Robots with LEGO MINDSTORMS* (Syngress Publishing, ISBN: 1-928994-67-9). Mario estimates he owns over 60,000 LEGO pieces. Mario works in Modena, Italy, where he lives with his wife, Anna, and his children, Sebastiano and Camilla.

Giulio Ferrari is a student in economics at the University of Modena and Reggio Emilia, where he also studied engineering. He is fond of computers and has developed utilities, entertainment software, and Web applications for several companies. Giulio discovered robotics in 1998, with the arrival of MIND-STORMS, and held an important place in the creation of the Italian LEGO community. He shares a love for LEGO bricks with his oldest brother, Mario, and a strong curiosity for the physical and mathematical sciences. Giulio also has a collection of 1200 dice, including odd-faced dice and game dice. Giulio has contributed to two other books for Syngress Publishing, *Building Robots with LEGO MINDSTORMS* (ISBN: 1-928994-67-9) and *Programming LEGO MINDSTORMS with Java* (ISBN: 1-928994-55-5). Guilio studies, works, and lives in Modena, Italy.

About This Book

Each of the ten cool robots in this book is presented using a method that makes its construction as easy and intuitive as possible. Each chapter begins with a picture of the completed robot, accompanied by a brief introduction to the robot's history, its unique challenges and characteristics, as well as any concerns that the robot's creator wants you to be aware of during construction.

The instructions for building each robot are broken down into several sub-assemblies, which each consist of an integral structural component of the finished robot. (For example, the first robot presented in this book, *WideBot*, is broken down into four sub-assemblies: the Chassis, the Right Drive, the Left Drive, and the Head.) You will see a picture of each finished sub-assembly before you begin its construction.

You will be guided through the construction of each sub-assembly by following the individual building steps, beginning with Step 0. Each step shows you two important things–what parts you need, and what to do with them–by using two pictures. The *parts list* picture shows you which LEGO bricks you will need for that particular step, as well as the quantity of parts required, and the color of the parts (if necessary). Since this book is printed in black and white, we have used the following key to represent the colors:

- **B** Blue
- **G** Green
- **M** Magenta
- **LB** Light Blue
- **Y** Yellow
- **Ppl** Purple
- **TLG** Transparent Light Green
- **TY** Transparent Yellow

The *instructional* picture next to the parts list shows how those parts connect to one another. As the robot's construction progresses, it gets harder to see where parts get added, so you'll see we have made the parts that you add in each particular step *darker than* those added in previous steps. Many of the steps also have a few brief lines of text to more fully explain building procedures that may not be obvious from the pictures alone, or to discuss what role this step plays in the larger scheme of the robot's construction.

Once you have finished building all of the separate sub-assemblies, it's time to put them all together to complete the robot. The set of steps at the end of each chapter titled "Putting It All Together" walks you through the process of attaching together the sub-assemblies.

Throughout the chapters you will see three types of sidebars:

- **Bricks & Chips…** These sidebars explain key LEGO building concepts and terminology.
- **Developing & Deploying…** These sidebars explain why certain building techniques were used with a particular robot and what purpose they serve.
- **Inventing…** These sidebars offer suggestions for customizing the robots. Building robots with LEGO MINDSTORMS is all about creativity, so we encourage you to experiment with these suggestions, and try different building techniques of your own.

Building your robots is, or course, only half the fun! Getting them to run using the RCX brick is what distinguishes MINDSTORMS robots from ordinary models created with LEGO bricks. Some of the robots in this book will use the programs that come hard-wired into the RCX brick. Many of them will use unique programs that the authors have written specifically for their robots. Keep an eye out for the black and white *syngress.com* icons scattered throughout the book.

These icons alert you to the fact that there is code for this particular robot available for download from the Syngress Solutions Web site (www.syngress.com/solutions). The programs for the robots in this book are written in two of the most common programming languages used for LEGO MINDSTORMS:

- **RCX** LEGO's official programming language.
- **NQC** Standing for "Not Quite C," NQC is a programming language created by Dave Baum. Very similar in many ways to the C computer programming language, NQC is a text-based language that is more powerful and flexible than RCX.

For instruction on uploading these programs to your RCX brick, refer to the documentation that came with your LEGO MINDSTORMS RIS 2.0 kit.

The Syngress Solutions Web site (www.syngress.com/solutions) contains the code files for the robots found in *10 Cool LEGO Mindstorms Ultimate Builders Set Projects: Amazing Projects You Can Build in Under an Hour.* The code files are located in a *BotXX* directory. For example, the files for Robot 5 are located in folder Bot05. Any further directory structure depends upon the specific files included for the robot in that particular chapter.

Contents

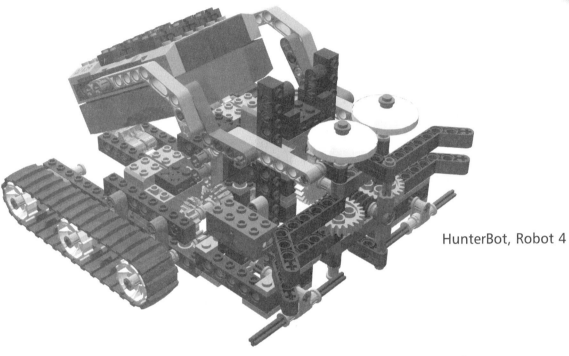

HunterBot, Robot 4

Foreword

My contribution to this book is the result of a series of accidents. First, my parents accidentally started me on the road to LEGO addiction by giving me countless LEGO kits as a child, which I played with obsessively. Later in my life, my little brother accidentally left his LEGO kits unguarded, and I rediscovered LEGO even though I was obviously older than the implied "9+" age category on the package. The final accident that led to my involvement with this book occurred in March 2001 when a co-worker–a programmer–told me about the LEGO MINDSTORMS Droid Developer Kit he had recently purchased. That triggered my relapse into LEGO addiction, and I had to buy the LEGO MINDSTORMS Robotics Invention System. Since then, I have been happily designing, building, and programming dozens of robots.

From time to time, I browse the web to check out the latest inventions by other MINDSTORMS builders, and I'm often impressed with the creativity and mechanical complexity that I find. I'm also amazed–and oddly reassured–by how many MINDSTORMS fanatics are adults who have picked up a fascinating new hobby.

In my first weeks working with the Robotics Invention System, I was a bit daunted by the robotic masterpieces that I saw on the web: mechanical arms, climbers, quadrupeds, bipeds, insects, cars, and even kitchen appliances. I wondered if I could ever begin to invent such things.

The key, I think, is to start slowly and innovate on existing models. Imitation is the sincerest form of flattery, but it also helps you learn the ropes. Spend some time browsing the web for ideas and see if you can innovate on them. The projects presented in the LEGO MINDSTORMS Ultimate Builders Set are an excellent way to pick up new building techniques and learn the principles of reinforced structures and reduction gearing. With dedication, anyone can learn to invent robots like a LEGO Master Builder.

When I was approached by Jonathan Babcock at Syngress to design robots for this book, I was thrilled to have the opportunity to contribute some of my ideas. An important aspect of the MINDSTORMS community, I feel, is the sharing of ideas and knowledge for the sake of fun and innovation. This book contributes to that community in the same spirit. I hope that you enjoy the cool and easy-to-build robots in this book, and use them as a springboard to designing your own innovative MINDSTORMS creations.

–Stephen Cavers
www.cavers.ca

Robot 1

WideBot

WideBot is a wide-bodied robot designed to pursue light sources. When I bought my second RIS kit, I tried to think up ways to get two robots to interact with each other. After a bit of experimentation, I put together a game of robotic tag called *Cat and Mouse*.

In this game, two robots sit still most of the time, but at random intervals, the "cat" robot sends a message asking if the "mouse" robot wants to play. If it does, the mouse wanders around the floor, evading the cat.

The mouse robot has a bright tail light that shows up very well to the cat robot's light sensor. The cat scans left and right, searching for the brightest point in the room, which it pursues. When the cat tags the mouse, it sends an infrared message to shut down the mouse. Then they wait quietly until they feel like playing again.

The WideBot is based upon my design for the cat. Your "mouse" equivalent can be a flashlight or other source of bright illumination that you can direct. There are several strategies to programming a light-seeking robot. You might wish to explore some of these as you experiment with WideBot's abilities. One strategy is to rotate the sensor (or the entire robot) from side to side in a sensor sweep, while recording the brightness values at several points across the sweep. The robot then determines the brightest direction and moves in that direction for a distance before performing the sensor sweep again. While this technique is probably an accurate way to locate the brightest point in a room, it's difficult to program properly.

A simpler way to seek for a light source is for the robot to turn steadily in one direction until the light level exceeds a specified threshold. This behavior is much easier to program, but doesn't take variable lighting conditions into account (if a room is too dark, for example, the robot may never register a light source bright enough to pursue).

WideBot uses the latter method, the same principle as the cat in the robotic tag game. By rotating on its wide wheelbase, WideBot can carefully scan a room for the light source. WideBot's width isn't just for looks: The wide turning circle makes a side-to-side scanning motion that is slow and easily controlled.

You can find the program for WideBot on the Syngress Solution Web site (www.syngress.com/solutions). It is a simple light-seeking program. The robot will keep turning in one direction until either the light exceeds the *LIGHT_THRESHOLD* value or until one of the bumpers is hit. The direction the robot turns is determined by which bumper is hit (if the left bumper is hit, the robot seeks to the right and vice-versa). The robot moves straight ahead while the light is above the threshold.

The Chassis

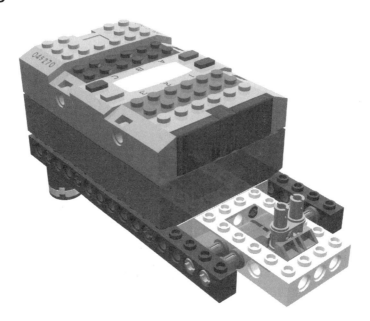

The chassis sub-assembly is the frame on which WideBot is built. It supports the RCX brick and provides a solid connection for the other parts. Using the open-centered transparent bricks, you can quickly build WideBot's sturdy frame.

Chassis Step 0

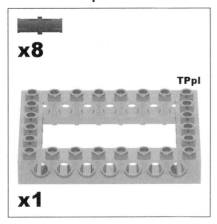

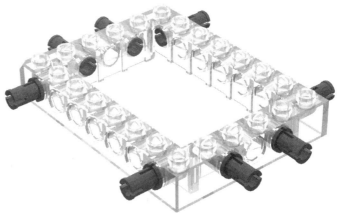

Add the pins to the 6x8 open-centered transparent brick.

Chassis Step 1

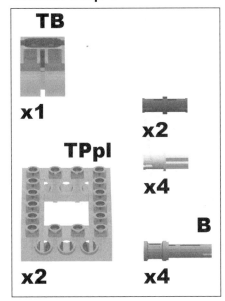

TB

x1

x2

TPpl

x4

B

x2

x4

Add the two 4x6 open-centered bricks. The transparent blue connector block will ultimately support WideBot's head.

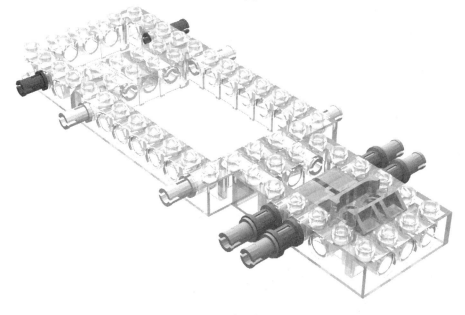

Chassis Step 2

x2

Y

x1

Attach the two 1x16 TECHNIC beams, which reinforce the chassis and provide attachment points for the left and right drive sub-assemblies.

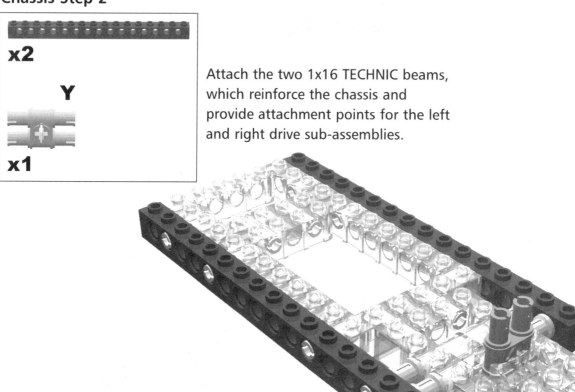

Chassis Step 3

x2

x2

x2

Assemble and attach the skids. Because WideBot is a two-wheeled robot, it needs skids in order to turn freely.

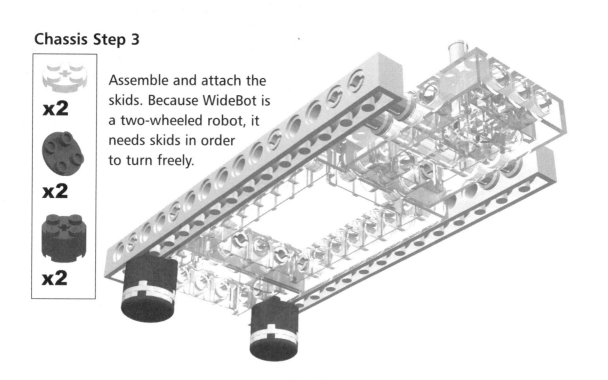

Inventing…

Casters, Skids, or Sliding?

Because WideBot uses skids it runs best on a smooth surface. If a smooth surface is not available you could try to adapt the caster wheel that appears in Robot 5, Nessie, or the sliding wheel that appears in Robot 6, Nellie.

Chassis Step 4

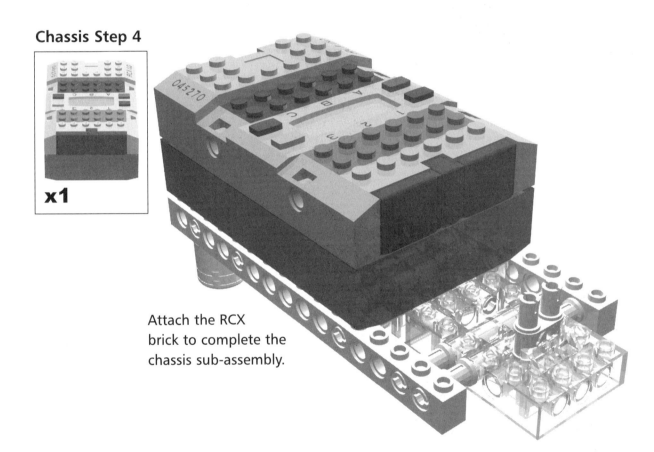

x1

Attach the RCX brick to complete the chassis sub-assembly.

The Right Drive

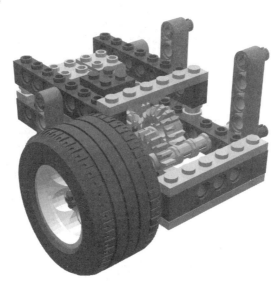

The right drive sub-assembly is a complete drive unit that includes the motor, gearbox, and right wheel. The gearbox turns the rotation angle 90 degrees and reduces the rotation speed by a ratio of 5:1. In other words, for every five revolutions of the motor, the wheel turns only once.

WideBot has two drive sub-assemblies, the right and the left, which include connectors and supports for the chassis.

Right Drive Step 0

Attach the 1x12 TECHNIC beam to the motor using the 1x2 plates, and slide the 12t beveled gear onto the motor.

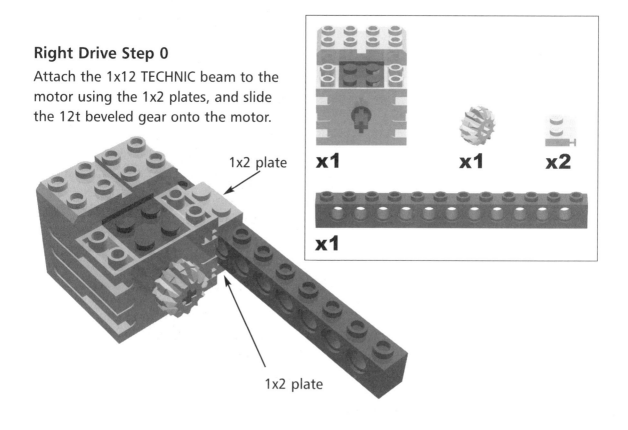

1x2 plate

1x2 plate

x1   x1   x2

x1

Right Drive Step 1

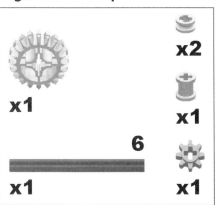

x1

6

x1

x2

x1

x1

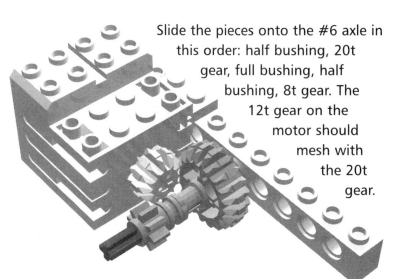

Slide the pieces onto the #6 axle in this order: half bushing, 20t gear, full bushing, half bushing, 8t gear. The 12t gear on the motor should mesh with the 20t gear.

Bricks & Chips…

Beveled Gears

The Ultimate Builders Set includes three kinds of beveled gears, which allow you to turn the angle of rotation 90 degrees.

Right Drive Step 2

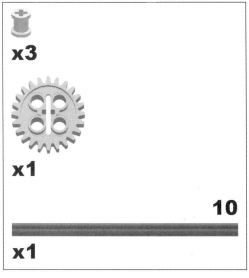

x3

x1

10

x1

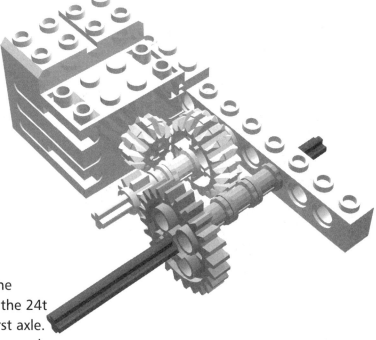

Slide the 24t gear and bushings onto the second axle, and place the axle so that the 24t gear meshes with the 8t gear on the first axle. These gears further reduce the rotation speed.

Right Drive Step 3

Slide the second 1x12 TECHNIC beam onto the axles, and secure it to the motor using the 1x2 plate.

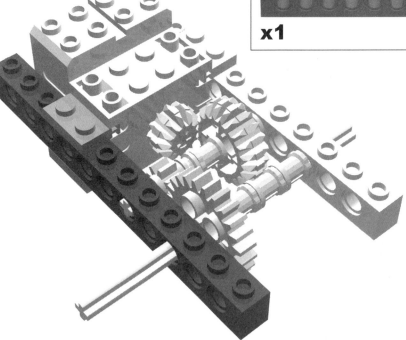

Right Drive Step 4

Add the various plates and bushings as shown. The 1x4 TECHNIC brick provides more strength to the assembly.

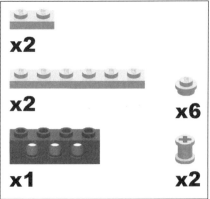

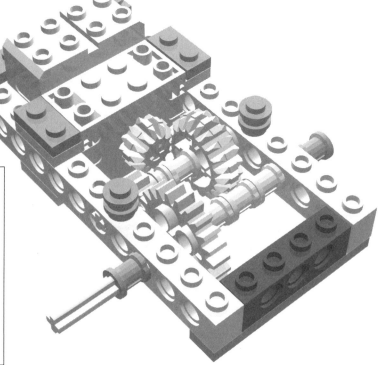

Right Drive Step 5

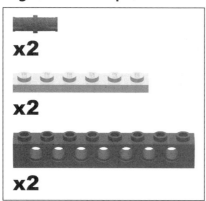

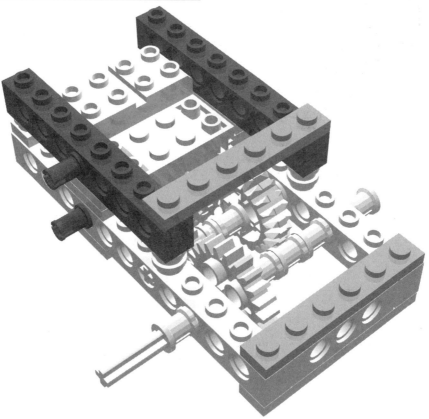

Bricks & Chips…

Using 1x1 Round Plates as Spacers

Models in many MINDSTORMS and TECHNIC kits stack two
1x1 plates to separate beams. This ensures that the holes line
up when you add a vertical beam, and is a lighter and more
attractive solution than filling the gap with long plates.

Right Drive Step 6

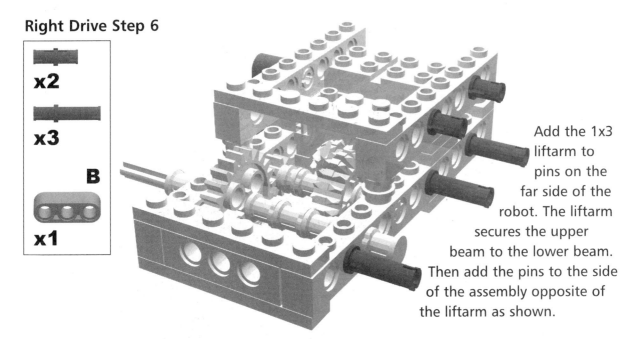

Add the 1x3 liftarm to pins on the far side of the robot. The liftarm secures the upper beam to the lower beam. Then add the pins to the side of the assembly opposite of the liftarm as shown.

Right Drive Step 7

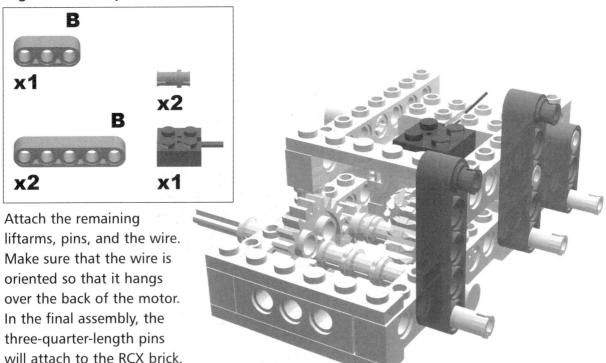

Attach the remaining liftarms, pins, and the wire. Make sure that the wire is oriented so that it hangs over the back of the motor. In the final assembly, the three-quarter-length pins will attach to the RCX brick.

Right Drive Step 8

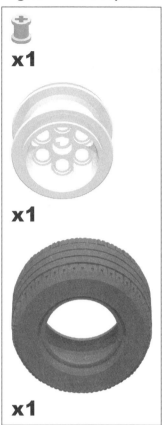

x1

x1

x1

Finally, slide the wheel onto the forward axle and fix it in place with a bushing. Now you're ready to build the left drive sub-assembly.

The Left Drive

The left drive sub-assembly is a mirror image of the right drive sub-assembly, with the exception of the gearbox. To ensure that the wheels drive in the same direction, the gearbox setup is identical in both motor assemblies (that is, it is not a mirror of the right drive, but is exactly the same).

Left Drive Step 0

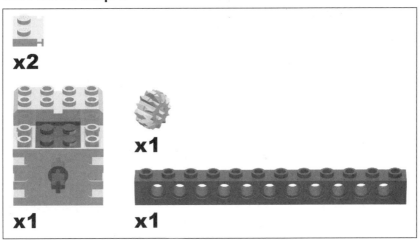

x2

x1

x1

x1

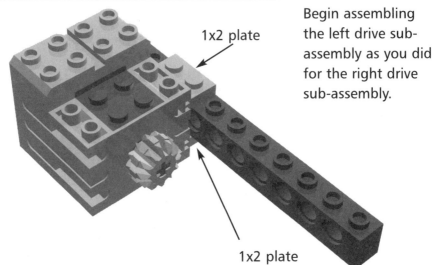

1x2 plate

Begin assembling
the left drive sub-
assembly as you did
for the right drive
sub-assembly.

1x2 plate

Left Drive Step 1

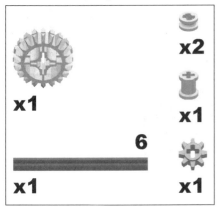

x1

x2

x1

6

x1

x1

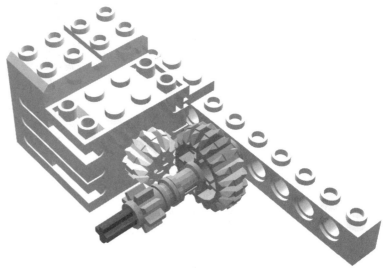

Left Drive Step 2

Here is the first difference between the two drive sub-assemblies: The axle extends out to the left instead of the right.

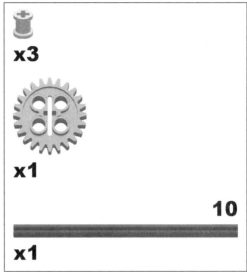

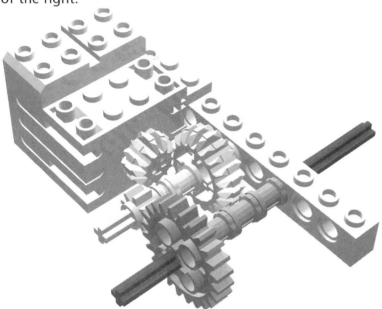

Left Drive Step 3

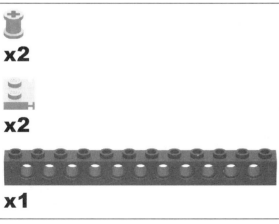

Left Drive Step 4

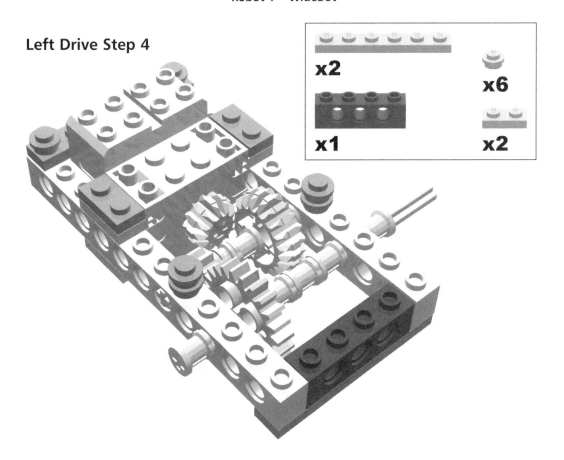

Left Drive Step 5

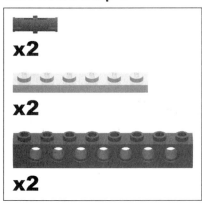

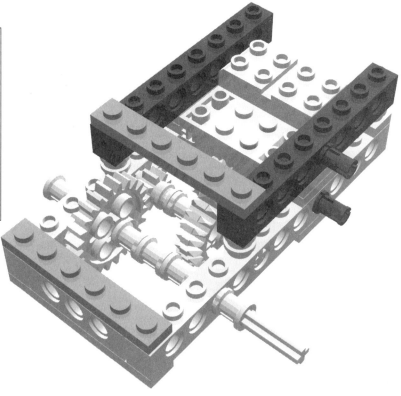

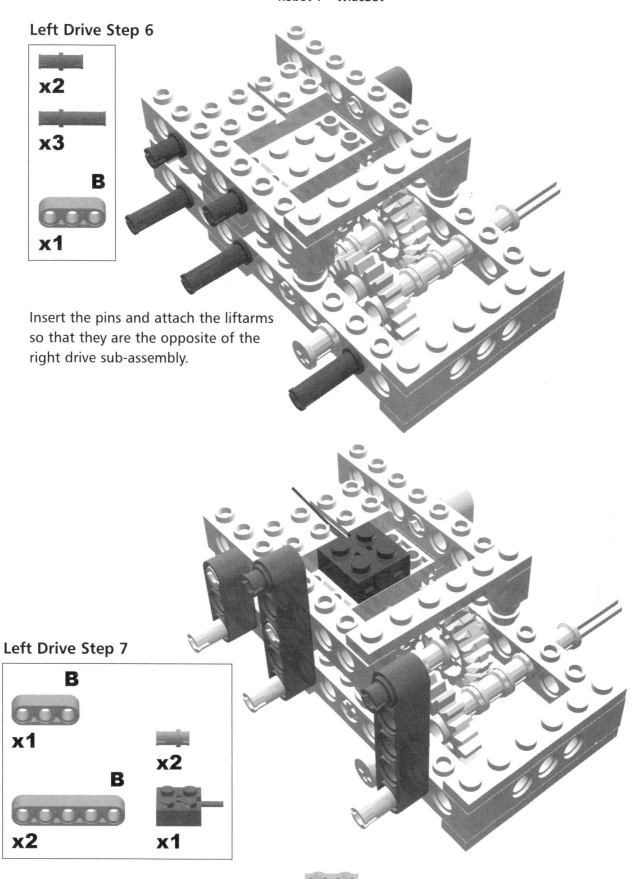

Left Drive Step 6

x2

x3

B

x1

Insert the pins and attach the liftarms
so that they are the opposite of the
right drive sub-assembly.

Left Drive Step 7

B

x1

B

x2

x2

x1

Left Drive Step 8

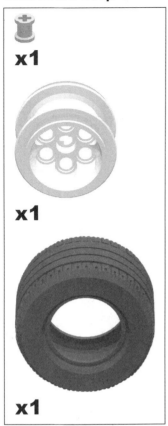

Complete the sub-assembly by sliding the left wheel onto the axle with a bushing. Now that the drive sub-assemblies are done, you can go ahead and build WideBot's head.

The Head

The head sub-assembly provides WideBot with two senses: touch and sensitivity to light. With touch sensors behind the left and right bumpers, WideBot can respond appropriately when it bumps into an object. The light sensor measures brightness directly ahead.

Developing & Deploying…

Using a Light Sensor as a Proximity Sensor

A forward-pointing light sensor can be used as a proximity detector. As the sensor approaches an object, the red light from the sensor reflects back to the sensor, much like radar.

Head Step 0

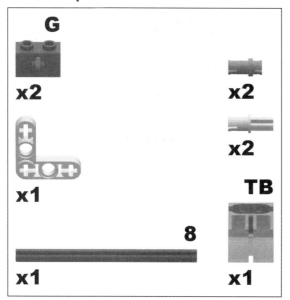

Head Step 1

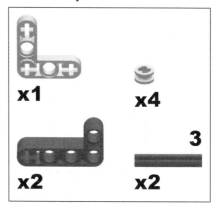

Assemble the parts on the transparent blue connector block. The green bricks with axle holes are the attachment point for the light sensor.

Slide the 3L liftarms and half-bushings onto the upper axle.

Inventing…

Heads Up or Heads Down?

If you want to build a line-following robot instead of a light-following robot, rotate the bricks with axle holes 90 degrees forward. The light sensor can then point downward instead of forward.

Head Step 3

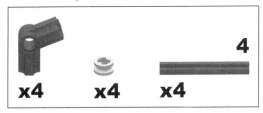

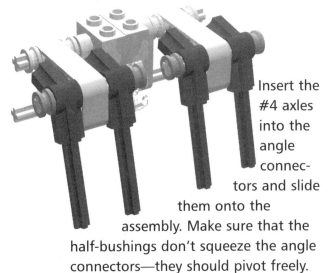

Insert the #4 axles into the angle connectors and slide them onto the assembly. Make sure that the half-bushings don't squeeze the angle connectors—they should pivot freely.

Head Step 2

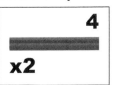

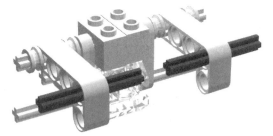

Slide the axles into the liftarms.

Head Step 4

Slide the 1x3 liftarms onto the axles.

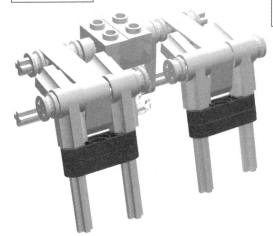

Head Step 5

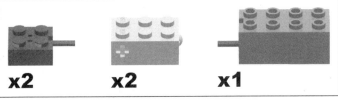

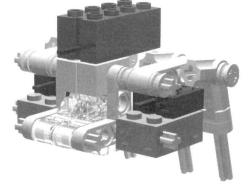

Slide the two touch sensors onto the axles, and attach the light sensor to the green bricks so that it faces forward.

Head Step 6

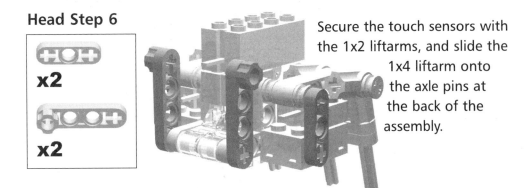

Secure the touch sensors with the 1x2 liftarms, and slide the 1x4 liftarm onto the axle pins at the back of the assembly.

Head Step 7

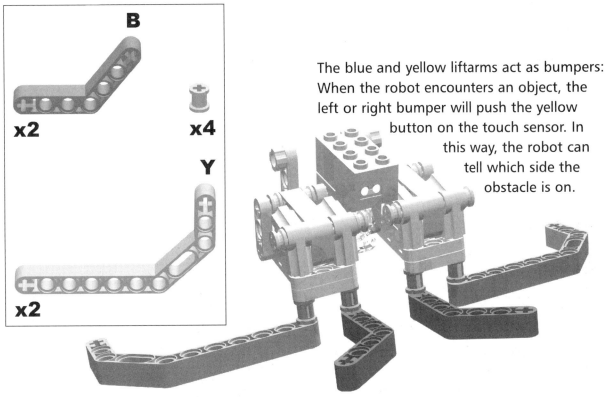

The blue and yellow liftarms act as bumpers: When the robot encounters an object, the left or right bumper will push the yellow button on the touch sensor. In this way, the robot can tell which side the obstacle is on.

Inventing…

Customizing the Bumpers

WideBot uses the blue and yellow liftarms as bumpers, but you can use other parts, such as flexible tubes or a combination of axles and connectors.

Head Step 8

2

x2

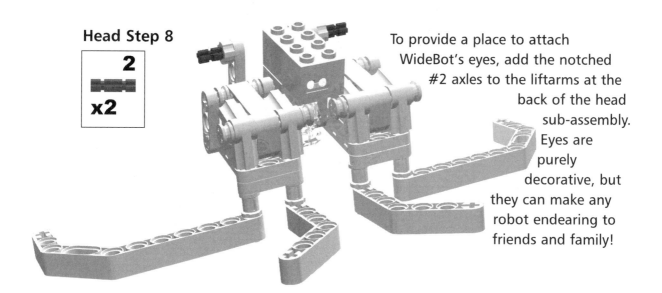

To provide a place to attach WideBot's eyes, add the notched #2 axles to the liftarms at the back of the head sub-assembly. Eyes are purely decorative, but they can make any robot endearing to friends and family!

Head Step 9

x2

2

x2

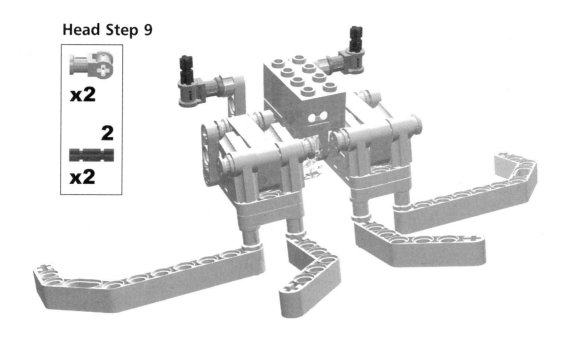

Head Step 10

B

x2

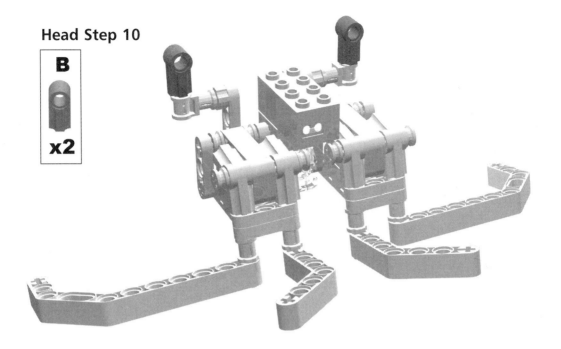

Head Step 11

B

x2

x2

Complete the head sub-assembly by attaching the eyes.

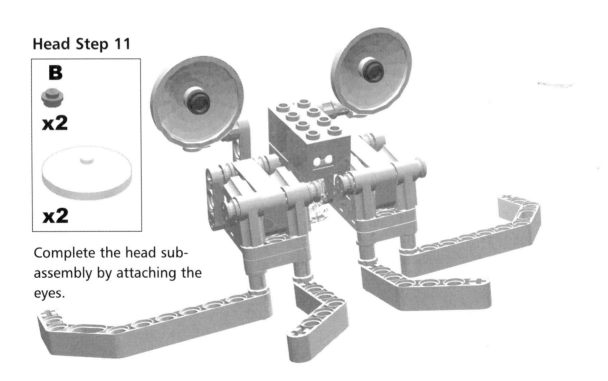

Putting It All Together

You're almost there. Now that you've built the chassis, head, and left and right drive sub-assemblies, you can put them together to complete WideBot.

Final Step 0

To begin the final assembly, start with the chassis sub-assembly. Attach the right drive sub-assembly to the right side of the chassis as shown.

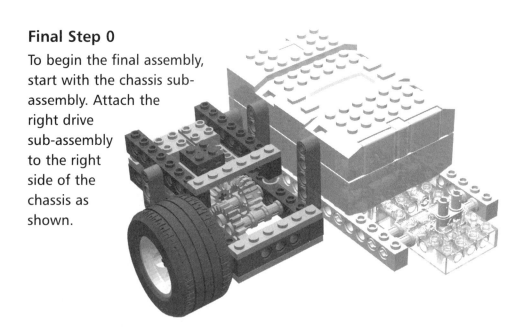

Final Step 1

Now, attach the left drive sub-assembly the same way to the left side of the chassis.

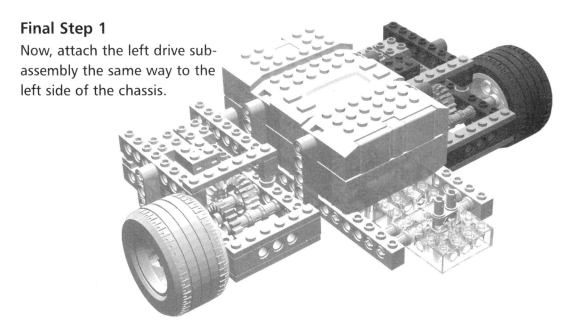

Bricks & Chips...

Changing the Batteries

To change the batteries, remove both drive sub-assemblies and separate the RCX brick from the chassis.

Final Step 2

Snap the head sub-assembly onto the yellow double pin at the front of the chassis.

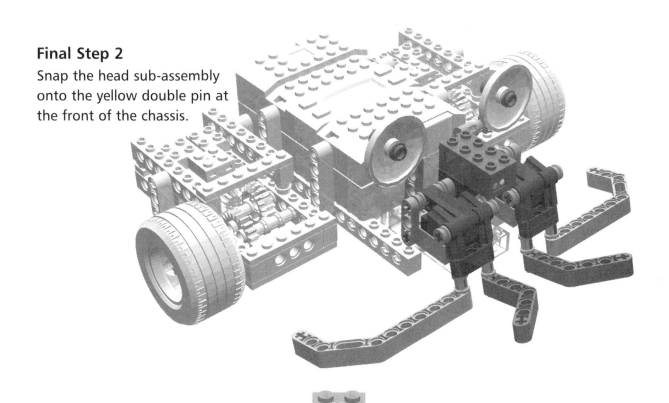

Final Step 3

x5

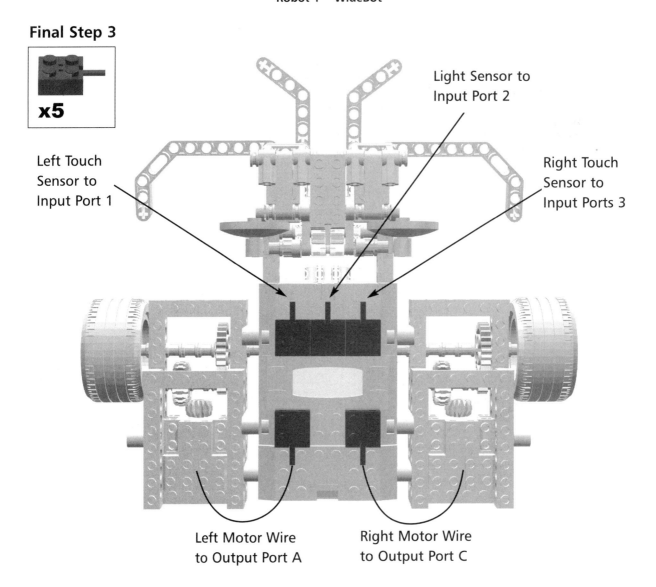

Left Touch
Sensor to
Input Port 1

Light Sensor to
Input Port 2

Right Touch
Sensor to
Input Ports 3

Left Motor Wire
to Output Port A

Right Motor Wire
to Output Port C

Now attach the wires.

Connect the left and right touch sensors to Input Ports 1 and 3, and connect the light sensor to Input Port 2.

Connect the left motor wire to Output Port A and the right motor wire to Output Port C. Be sure to orient the motor wires as shown: The wires should hang over the back of the RCX brick. WideBot should move forward when both motors run in the forward direction.

WideBot is now complete! Load the program and have some light-detecting, obstacle-bumping fun.

SYNGRESS
syngress.com

Developing & Deploying...

A Light-following Program

You can download the program for WideBot from the Syngress Solutions Web site (www.syngress.com/solutions).

Robot 2

SumoBug

One of the popular pastimes of LEGO MINDSTORMS builders is a challenge called *Robotic Sumo*, which emulates Sumo wrestling. In these challenges, two robots face each other in a ring and employ a combination of cunning and brute force in a bid to shove their opponent out of the ring. With the Robotics Invention System (RIS) Test Pad, you can stage your own Sumo matches within the Test Pad's large, black, oval "ring."

SumoBug is designed with the strategies of Sumo in mind. Although it's a bit of a lightweight in Sumo terms, SumoBug keeps its weight over the front axles for greater traction. The slow-turning tractor treads keep it stuck to the ground and moving forward.

SumoBug gets its power from a 24:1 gear ratio, which means that it moves slowly but has lots of torque. Although some Sumo robots try to crash head-long into an opponent, SumoBug's strategy is to stay low to the ground and push slowly and steadily to win the match.

SumoBug is versatile, too. If there are no opponents handy (or if you don't want to buy a second RIS just yet), you can program SumoBug to detect and follow lines on the floor and navigate obstacles—it works with almost any program for a robot with two bumpers and a light sensor, such as the programs that come with the RIS 2.0 software.

Here's how you can set up a Sumo match on your RIS Test Pad. Build two SumoBugs and download the SumoBug program from the Syngress Solutions Web site (www.syngress.com/solutions). The program is available in both NQC and RCX. The two versions are very similar, so you can use either version in a Sumo match. These programs allow two dueling Sumo robots to attempt to force each other out of the black oval on the LEGO MINDSTORMS Test Pad.

The SumoBug program requires that both opponents have the following characteristics:

- Left and right front bumpers with touch sensors attached to Input Ports 1 and 3

- A downward-pointing light sensor attached to Input Port 2 and mounted on the rear of the robot

- Left and right motors connected to Output Ports A and C, respectively

Place the two robots so that they are facing each other across the length of the Test Pad. Run the program to start the match. If your SumoBug is forced backward over the black line, it loses the match and sends a "you win" message to the victor. If your SumoBug receives a "you win" message before crossing the black line, it does a little victory dance and plays a tune.

Bricks & Chips...

SumoBug Is All Torque and No Action

By reducing the rotation speed, you increase the torque, which makes for a slow but very powerful robot. Finding the right gear ratio means balancing the need for power and the need for speed.

The Right Drive

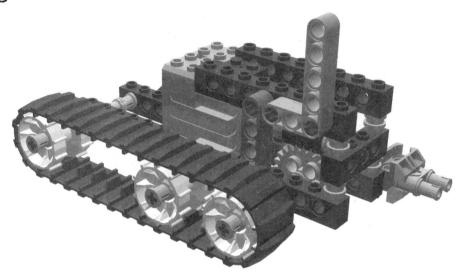

To begin, you will build the right drive sub-assembly, which includes the motor, gearbox, and right tractor tread. The gearbox uses a worm gear and a 24t gear to reduce the rotation speed by a ratio of 24:1. In other words, for every 24 revolutions of the motor, the wheel inside the tractor tread turns only once.

SumoBug's two drive sub-assemblies, the left and right drives, include connectors and supports for the RCX brick.

Right Drive Step 0

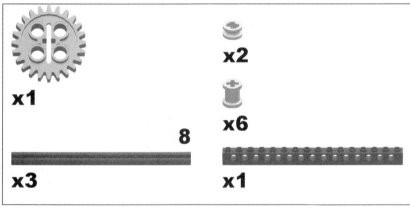

x1

x2

x6

8

x3

x1

Slide the axles through the holes in the 1x16 TECHNIC brick. Add the bushings and the 24t gear to hold the axles in place. The 24t gear is secured on the axle with the half-bushings.

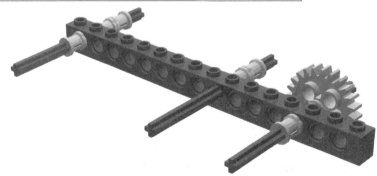

Right Drive Step 1

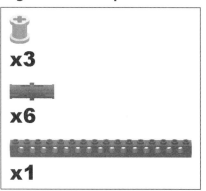

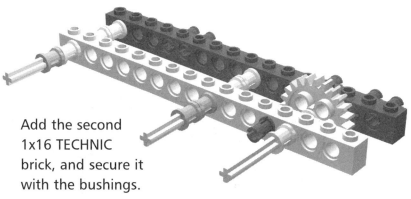

Add the second 1x16 TECHNIC brick, and secure it with the bushings.

Right Drive Step 2

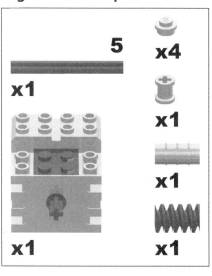

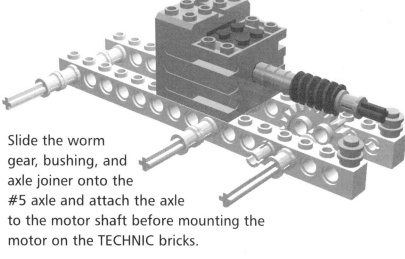

Slide the worm gear, bushing, and axle joiner onto the #5 axle and attach the axle to the motor shaft before mounting the motor on the TECHNIC bricks.

Right Drive Step 3

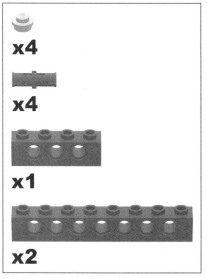

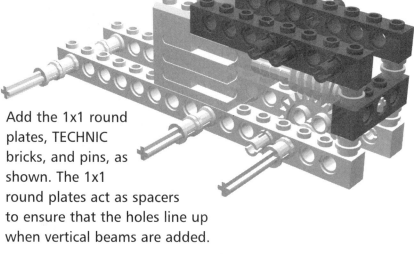

Add the 1x1 round plates, TECHNIC bricks, and pins, as shown. The 1x1 round plates act as spacers to ensure that the holes line up when vertical beams are added.

Right Drive Step 4

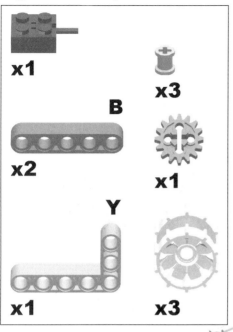

x1

x3

B

x2

x1

Y

x1

x3

Slide the 16t gear onto the front axle. Mount the sprocket wheels on to the axles with a bushing on each to hold them in place.

When you connect the wire to the motor, make sure that the orientation is correct: The wire should hang off the back of the motor.

The liftarms on each side strengthen the drive unit and provide the connection point for the RCX brick.

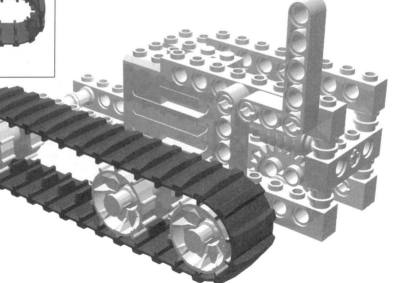

16t gear goes onto axle before wheel.

Right Drive Step 5

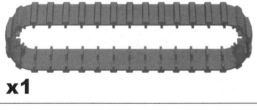

x1

Slide the tread over the sprocket wheels.

Right Drive Step 6

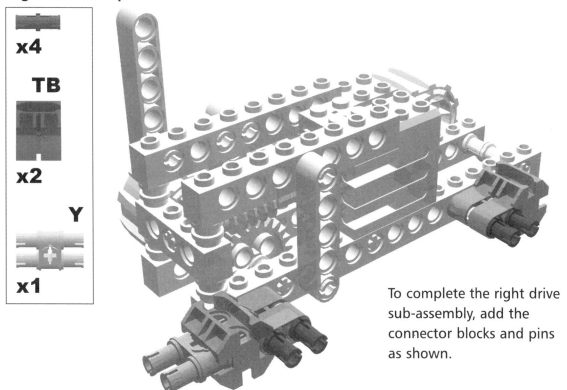

To complete the right drive sub-assembly, add the connector blocks and pins as shown.

The Left Drive

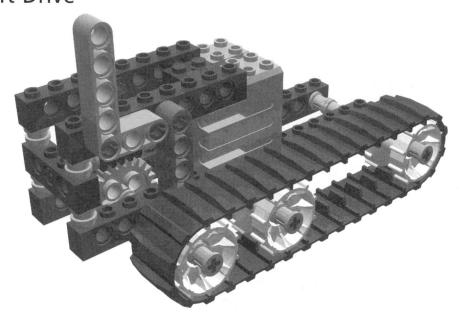

The left drive sub-assembly is almost a mirror image of the right drive sub-assembly, and includes the motor, gearbox, and tractor tread.

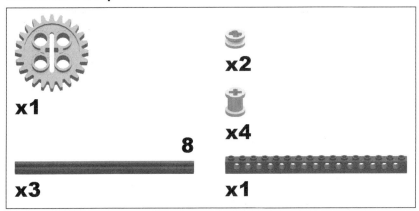

Bricks & Chips...

Saving Your Good Designs

It's a good idea to design self-contained drive sub-assemblies—you can easily reuse them in your future robot designs.

Left Drive Step 0

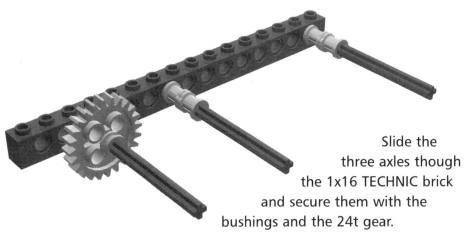

Slide the three axles though the 1x16 TECHNIC brick and secure them with the bushings and the 24t gear.

Left Drive Step 1

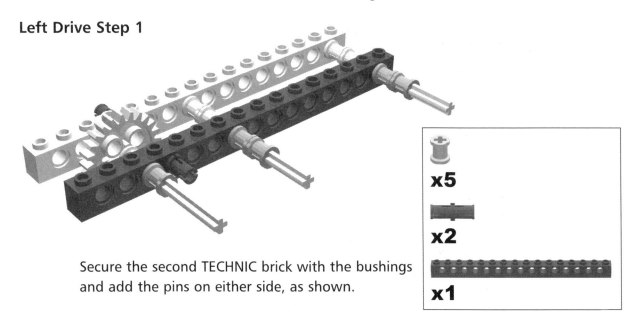

Secure the second TECHNIC brick with the bushings and add the pins on either side, as shown.

Left Drive Step 2

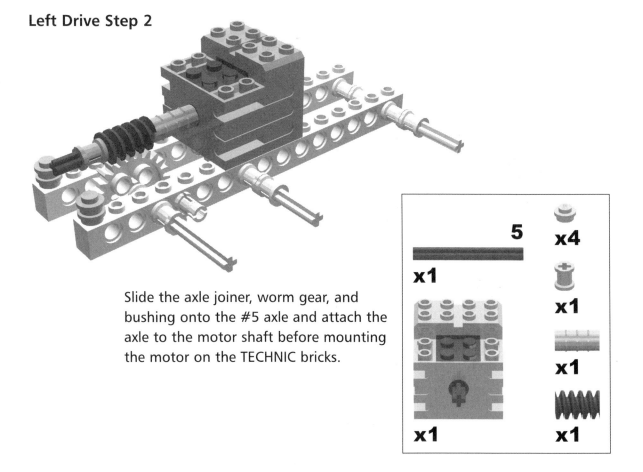

Slide the axle joiner, worm gear, and bushing onto the #5 axle and attach the axle to the motor shaft before mounting the motor on the TECHNIC bricks.

Left Drive Step 3

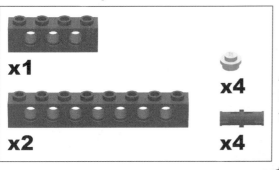

x1

x2

x4

x4

Add the spacers and TECHNIC bricks, which comprise the frame of the drive sub-assembly.

Left Drive Step 4

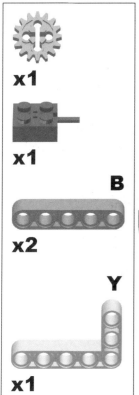

x1

x1

B

x2

Y

x1

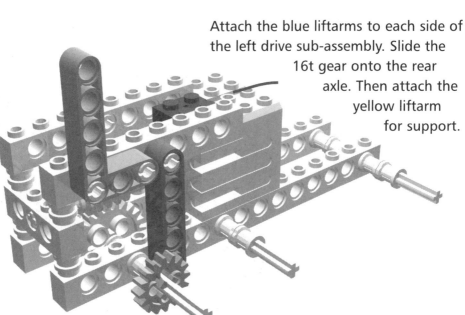

Attach the blue liftarms to each side of the left drive sub-assembly. Slide the 16t gear onto the rear axle. Then attach the yellow liftarm for support.

Left Drive Step 5

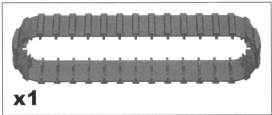

x3

x3

Attach the sprocket wheels to the three axles using the bushings, and slide the tractor tread onto the sprocket wheels.

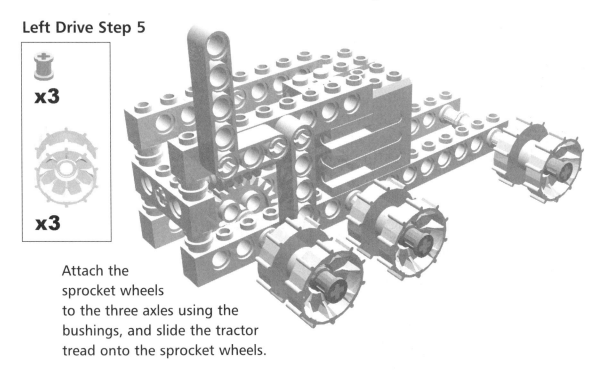

Left Drive Step 6

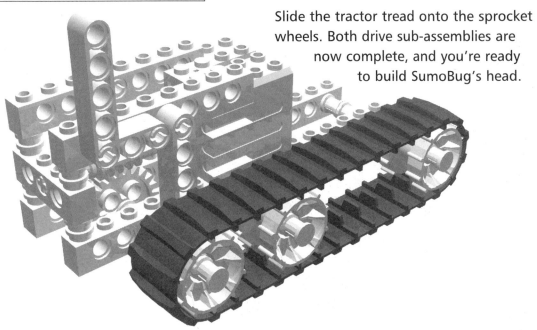

x1

Slide the tractor tread onto the sprocket wheels. Both drive sub-assemblies are now complete, and you're ready to build SumoBug's head.

The Head

SumoBug's head is more than just a pretty face. It holds the light sensor and two bumpers equipped with touch sensors. During a match, SumoBug uses the light sensor to see the black line on the RIS Test Pad. Separate bumpers let SumoBug know which side the opponent is on.

Head Step 0

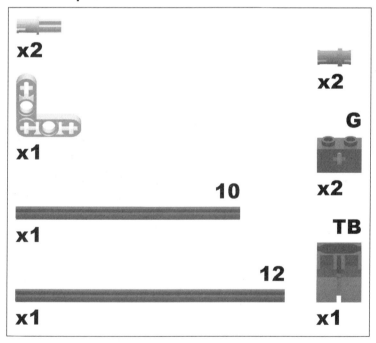

x2

x1

10

x1

12

x1

x2

G

x2

TB

x1

First, attach the pins to the transparent blue connector box. Then connect the liftarm. Slide the #10 axle through the center hole on the liftarm, and the #12 axle through the top hole. Next, add the green bricks with axle holes.

Bricks & Chips...

Keeping a Low Profile

An important strategy of Robotic Sumo is to stay close to the ground. If an opponent's robot gets underneath yours, it's game over. SumoBug's head is mounted close to the floor.

Head Step 1

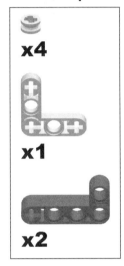

x4

x1

x2

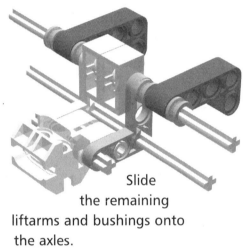

Slide the remaining liftarms and bushings onto the axles.

Head Step 2

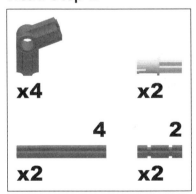

x4

x2

4

x2

2

x2

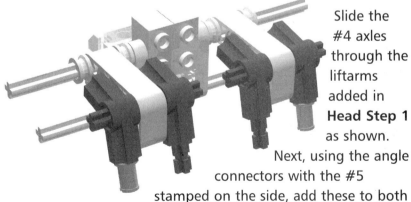

Slide the #4 axles through the liftarms added in **Head Step 1** as shown. Next, using the angle connectors with the #5 stamped on the side, add these to both ends of each axle. The angle connectors allow the front bumpers to pivot freely on the axles. Finally, add the pins to the exterior axle connectors on each side and the #2 axles to the interior axle connectors on each side.

Head Step 3

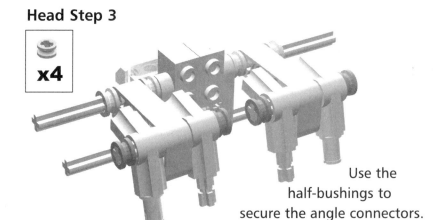

x4

Use the half-bushings to secure the angle connectors.

Head Step 4

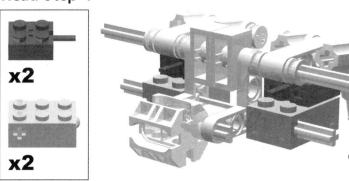

x2

x2

Connect a wire to each of the two touch sensors, and slide them onto the axle under the black liftarms. Once the bumpers are attached, the pivoting angle connectors will cause the bumper to push the touch sensor's yellow button when SumoBug encounters an obstacle.

Head Step 5

x2

x2

Use the 1x3 grey liftarms to secure the touch sensors on each side, and add the three-quarter-length pins.

Head Step 6

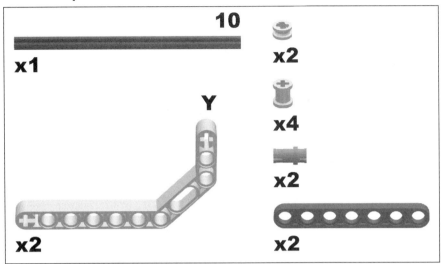

10
x1

Y

x2

x2

x4

x2

x2

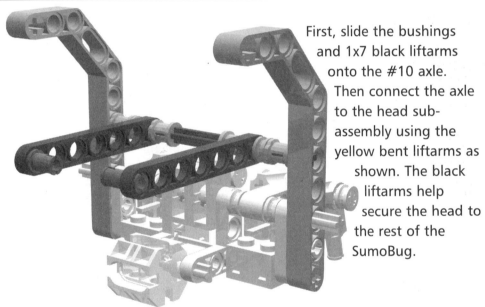

First, slide the bushings and 1x7 black liftarms onto the #10 axle. Then connect the axle to the head sub-assembly using the yellow bent liftarms as shown. The black liftarms help secure the head to the rest of the SumoBug.

Head Step 7

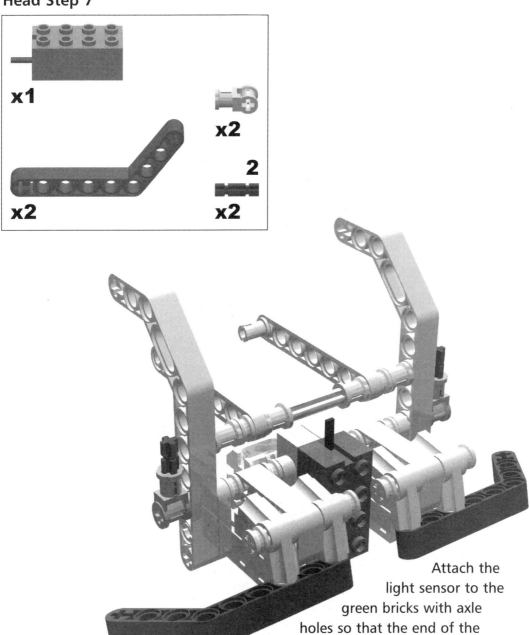

Attach the
light sensor to the
green bricks with axle
holes so that the end of the
light sensor extends slightly below the
bottom of the touch sensors.

Head Step 8

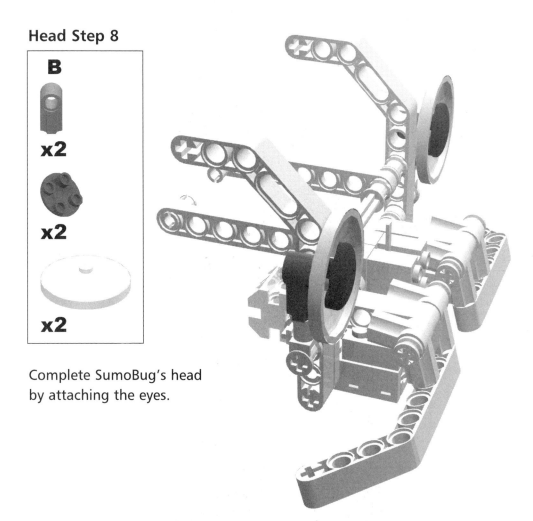

B

x2

x2

x2

Complete SumoBug's head by attaching the eyes.

Customizing SumoBug's Head

Although SumoBug's big, googly eyes are cute, not everyone will like them on their wrestler robot! Play around with different decorations, like antennae or smaller eyes to give your robot a different expression— or none at all!

The RCX

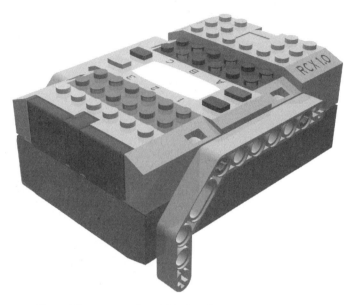

Next, you'll be attaching yellow liftarms to the RCX brick.

RCX Step 0

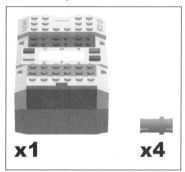

x1   **x4**

Insert three-quarter-length pins into each of the four holes on the RCX brick.

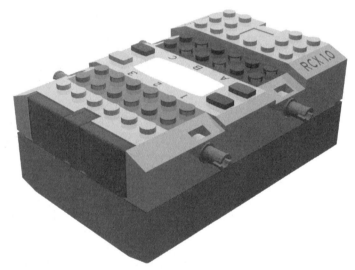

Bricks & Chips...

Looking Forward or Backward

Many MINDSTORMS builders orient the RCX brick so that the infrared window faces backward. Although partly an aesthetic issue, the important thing is to ensure that communication with the infrared tower is unobstructed.

RCX Step 1

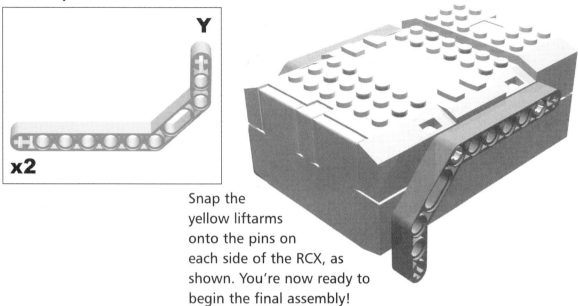

Snap the yellow liftarms onto the pins on each side of the RCX, as shown. You're now ready to begin the final assembly!

Putting It All Together

Now that you've completed all four major components–the left and right drive sub-assemblies, the head sub-assembly, and the RCX sub-assembly–you can go ahead and begin the final construction steps.

Bricks & Chips...

Where Is Your COG?

When traction is important, try to keep your robot's center of gravity (COG) over the drive wheels.

Final Step 0

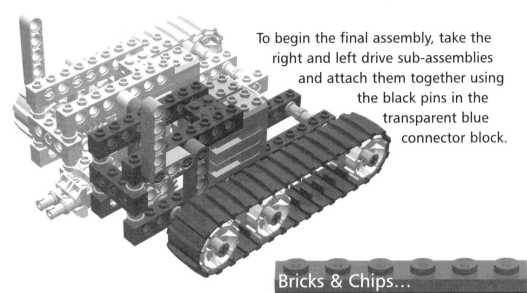

To begin the final assembly, take the right and left drive sub-assemblies and attach them together using the black pins in the transparent blue connector block.

Bricks & Chips...

Keeping Your Eyes on the Road

Notice how close the light sensor is to the ground. This lets the robot detect the brightness of the surface more accurately than if it were higher up.

Final Step 1

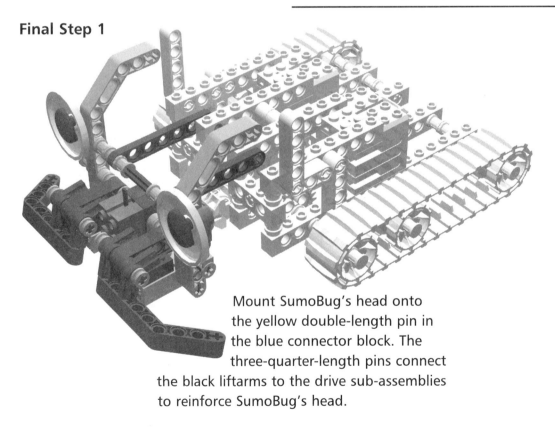

Mount SumoBug's head onto the yellow double-length pin in the blue connector block. The three-quarter-length pins connect the black liftarms to the drive sub-assemblies to reinforce SumoBug's head.

Final Step 2

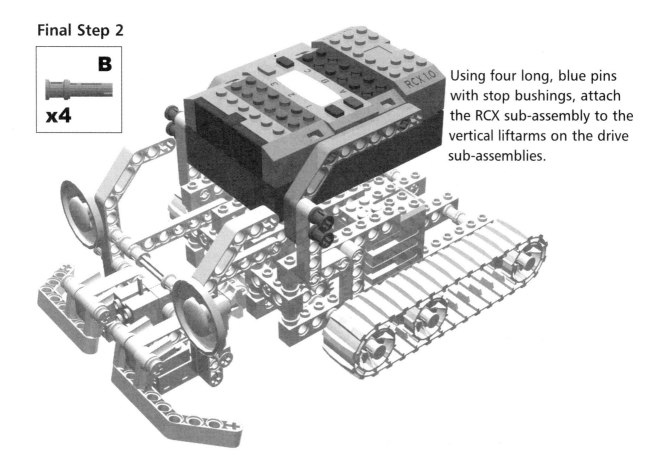

B

x4

Using four long, blue pins with stop bushings, attach the RCX sub-assembly to the vertical liftarms on the drive sub-assemblies.

Bricks & Chips...

Changing the Batteries

To change SumoBug's batteries, disconnect Output Ports A and C, then pull out the blue pins halfway to release the RCX brick.

Final Step 3

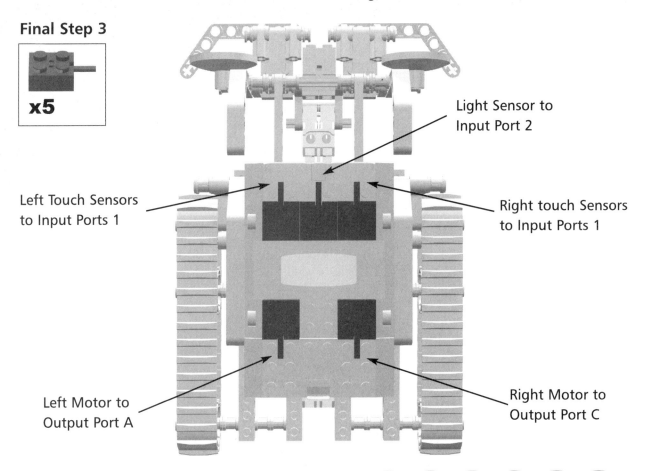

x5

Light Sensor to
Input Port 2

Left Touch Sensors
to Input Ports 1

Right touch Sensors
to Input Ports 1

Left Motor to
Output Port A

Right Motor to
Output Port C

Now it's time to connect the wires to the RCX brick. Connect the left and right touch sensors to Input Ports 1 and 3, and connect the light sensor to Input Port 2. Connect the left motor to Output Port A and the right motor to Output Port C, making sure that the wires are oriented as shown.

Developing & Deploying…
Programming SumoBug

The programs for SumoBug are located on the Syngress Solutions Web site (www.syngress.com/solutions). In addition to competing in robotic Sumo challenges, SumoBug can also be programmed as a line-following robot. Either type of program works with the RIS Test Pad.

Robot 3

Hopper

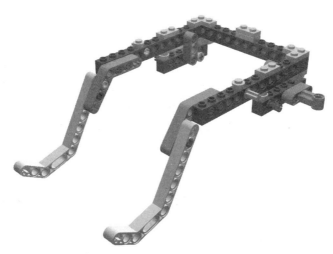

Hopper is a four-legged robot designed to move in a loping, hopping motion. Unlike most robots with legs, Hopper's design is extremely simple, which makes it an excellent starter project, and its unique gait makes it a fun robot to play with and program.

I spent quite a while trying to select an appropriate name for this robot. Having rejected such names as FleaBot, Skipper, and even Dennis (the Hopper), simplicity won out. Sometimes naming a robot can be the most difficult part of the invention process.

Hopper's hopping action is generated by the spinning action of the large pulley wheels, which drive the legs down and back, then retract them quickly. Although Hopper doesn't actually leave the ground on each hop, it does drive itself forward, and even has the ability to turn left and right. Positioning the weight correctly is crucial to achieve the desired hopping motion: Place the heavy RCX brick too far forward and there is too much weight on the front legs for Hopper to move; place it too far to the rear and the robot will flip onto its back.

Programming Hopper poses an interesting challenge. In order to hop forward, the legs need to move synchronously despite the fact that the motors run independently. You can synchronize the legs by simply allowing a small rest time between movements, during which the robot's own weight will return the legs to the resting state before starting movement again. If you're programming using the NQC (Not Quite C) programming language, use the *Float* command in place of the *Off* command. *Float* allows the motors to turn freely when they are turned off, which lets the legs return to the resting state in about half a second. If you program using the LEGO RCX language, allow at least a second or so between movements because the motors tend to resist turning.

SYNGRESS
syngress.com

The program for Hopper is designed to let it hop around the RIS Test Pad, following the black oval and hopping on the green rectangles. Due to the way the legs move, Hopper will only work on a smooth surface, so the Test Pad is an excellent place to let it roam. You can download the Hopper program from the Syngress Solutions Web site (www.solutions.com/syngress).

The Left Drive

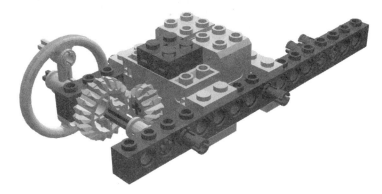

The left drive and right drive sub-assemblies are used to rotate the large pulley wheels which provide the hopping action for the legs. The beveled gears turn the angle of rotation 90 degrees and reduce the speed.

Left Drive Step 0

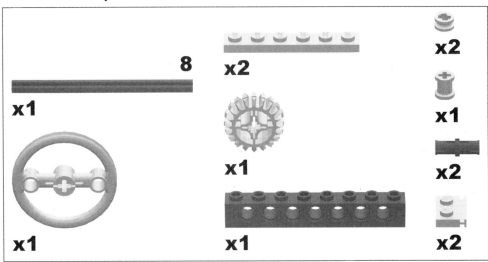

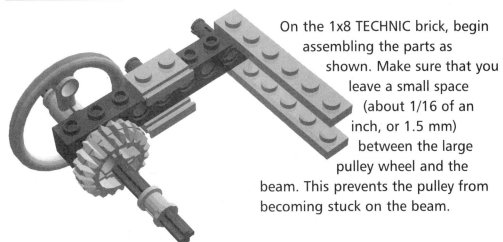

On the 1x8 TECHNIC brick, begin assembling the parts as shown. Make sure that you leave a small space (about 1/16 of an inch, or 1.5 mm) between the large pulley wheel and the beam. This prevents the pulley from becoming stuck on the beam.

Inventing...

Choosing the Gears

Your choice of gears determines speed and torque. You should experiment with different gear ratios—for example, if you replace Hopper's gears with an 8t gear and a 24t crown gear, Hopper will move more slowly.

Left Drive Step 1

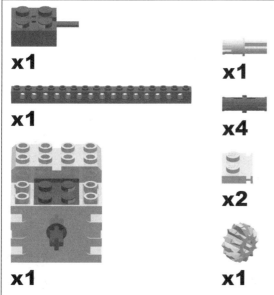

x1

x1

x1

x1

x4

x2

x1

Attach the 12t gear to the motor's drive axle, then slide the motor into place, so that the 12t gear meshes with the 20t gear. Attach the 1x16 beam.

Make sure that the wire is oriented correctly, as shown.

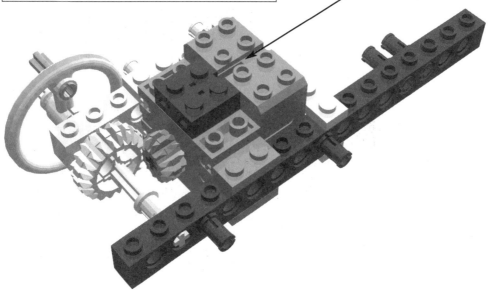

The Right Drive

The right drive sub-assembly is a mirror image of the left drive sub-assembly.

Right Drive Step 0

	8	**x2**
x1	**x2**	**x1**
	x1	**x2**
x1	**x1**	**x2**

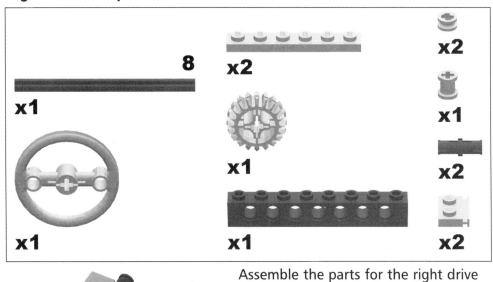

Assemble the parts for the right drive sub-assembly as you did for the left drive sub-assembly. Notice that the arrangement of gears is identical to that of the left drive sub-assembly. This ensures that the motors turn in the same direction.

Right Drive Step 1

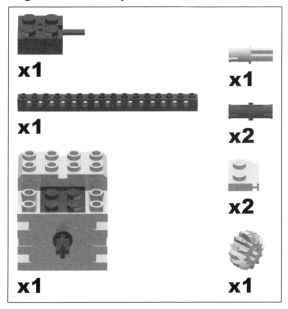

With the two motor sub-assemblies completed, you're ready to begin Hopper's chassis.

Add the remaining parts. Again, make sure that the wire is oriented as shown.

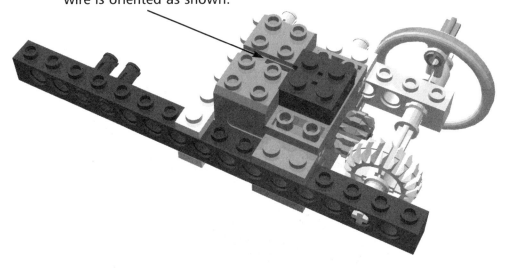

The Chassis

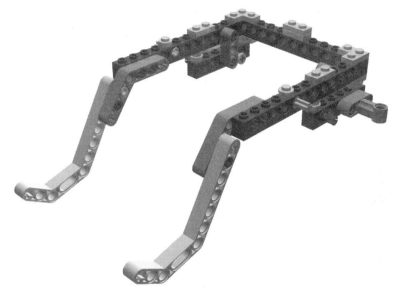

Hopper's chassis is a simple frame onto which of all of the other parts are added.

Chassis Step 0

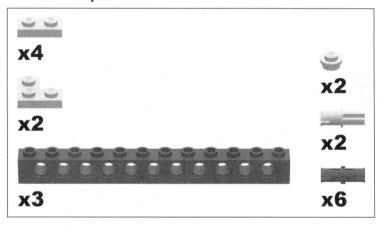

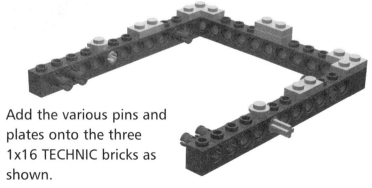

Add the various pins and plates onto the three 1x16 TECHNIC bricks as shown.

Chassis Step 1

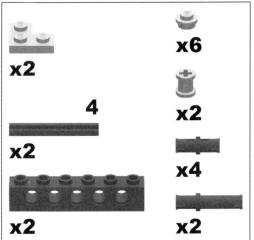

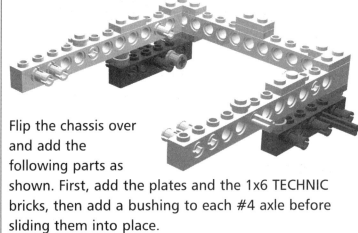

Flip the chassis over and add the following parts as shown. First, add the plates and the 1x6 TECHNIC bricks, then add a bushing to each #4 axle before sliding them into place.

Chassis Step 2

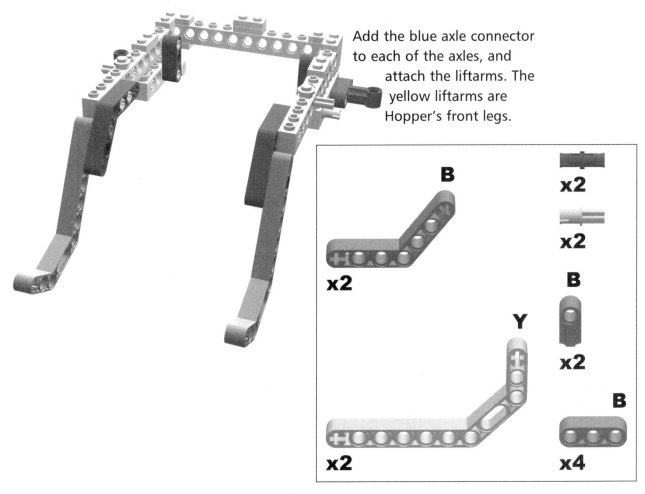

Add the blue axle connector to each of the axles, and attach the liftarms. The yellow liftarms are Hopper's front legs.

Bricks & Chips...

Using Skids Instead of Wheels

Hopper's front legs act as skids, which let the front end slide forward and side-to-side. When you use skids on a robot, avoid using rubber parts, such as tires, because they won't slide easily.

The RCX

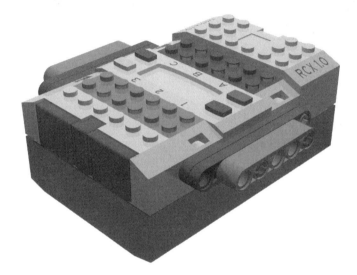

Before the RCX can be mounted onto Hopper's chassis, you need to add liftarms, which provide the attachment points.

RCX Step 0

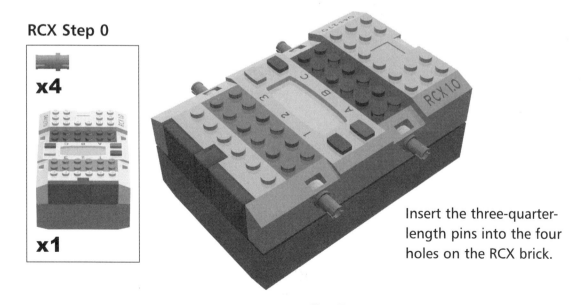

x4

x1

Insert the three-quarter-length pins into the four holes on the RCX brick.

RCX Step 1

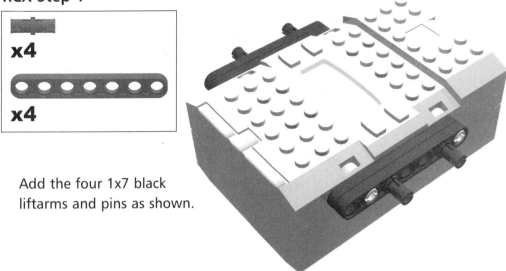

x4

x4

Add the four 1x7 black
liftarms and pins as shown.

RCX Step 2

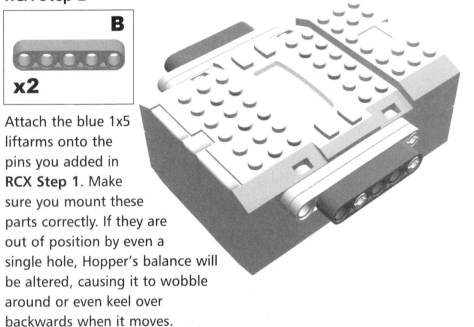

B

x2

Attach the blue 1x5
liftarms onto the
pins you added in
RCX Step 1. Make
sure you mount these
parts correctly. If they are
out of position by even a
single hole, Hopper's balance will
be altered, causing it to wobble
around or even keel over
backwards when it moves.

Inventing...

Choosing Part Colors

Your choice of part colors can influence the overall look of your robot.
If you prefer, you can use the blue 1x7 liftarms that came with the
Ultimate Builders Set. I chose the black ones simply for looks.

Putting It All Together

Now that you've built the left and right drive, chassis, and RCX sub-assemblies, you're ready to start Hopper's final assembly.

Final Step 0

To begin, locate the left drive sub-assembly and orient it as shown here.

Final Step 1

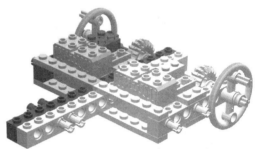

Locate the right drive sub-assembly and attach it to the left drive sub-assembly as shown.

Final Step 2

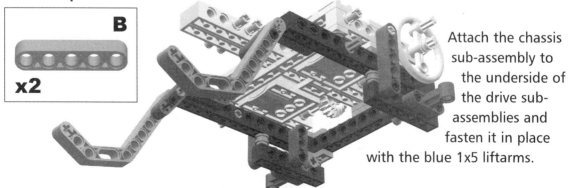

Attach the chassis sub-assembly to the underside of the drive sub-assemblies and fasten it in place with the blue 1x5 liftarms.

Final Step 3

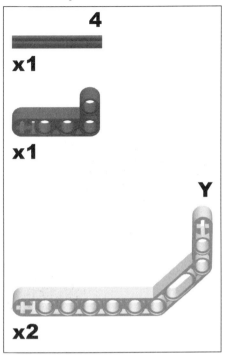

4
x1

x1

Y
x2

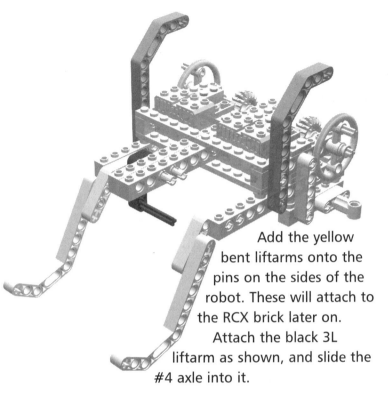

Add the yellow bent liftarms onto the pins on the sides of the robot. These will attach to the RCX brick later on. Attach the black 3L liftarm as shown, and slide the #4 axle into it.

Final Step 4

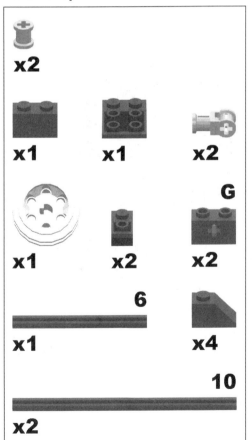

x2

x1   x1   x2

G
x1   x2   x2

6
x1   x4

10
x2

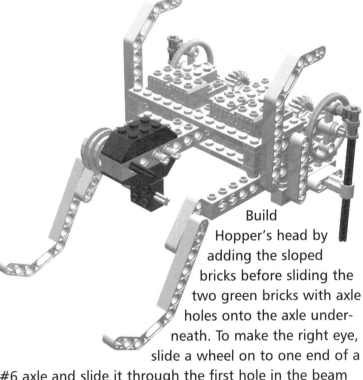

Build Hopper's head by adding the sloped bricks before sliding the two green bricks with axle holes onto the axle underneath. To make the right eye, slide a wheel on to one end of a #6 axle and slide it through the first hole in the beam as shown.

Using the #10 axles and bushings, build the legs on both sides of the robot.

Final Step 5

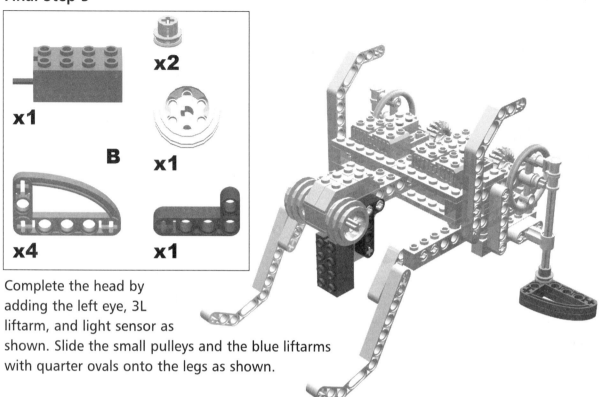

Complete the head by adding the left eye, 3L liftarm, and light sensor as shown. Slide the small pulleys and the blue liftarms with quarter ovals onto the legs as shown.

Inventing...

Starting Off on the Right Foot

Your choice of parts for the feet can change how Hopper moves. Try different parts in place of the ones shown to see what happens.

Final Step 6

Using the long blue pins with friction, fasten the RCX sub-assembly to the yellow liftarms as shown.

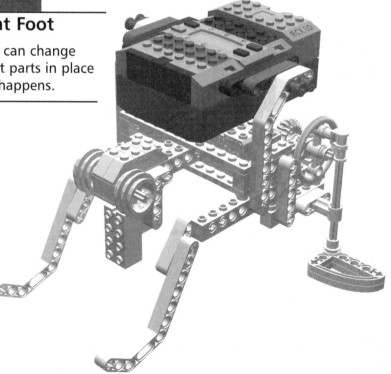

Bricks & Chips...

Changing the Batteries

To change the batteries, disconnect the wire from Input Port 2 and remove the blue pins.

Final Step 7

x3

Finally, connect the light sensor to Input Port 2, and connect the left and right motors to Output Ports A and C.

You're done! Once you download the program, you're ready get hopping.

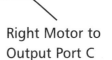

Light Sensor
to Input Port 2

Left Motor to
Output Port A

Right Motor to
Output Port C

Developing & Deploying...

Programming Hopper

You can download Hopper's program from the Syngress Solutions Web site (www.syngress.com/solutions).

Robot 4

HunterBot

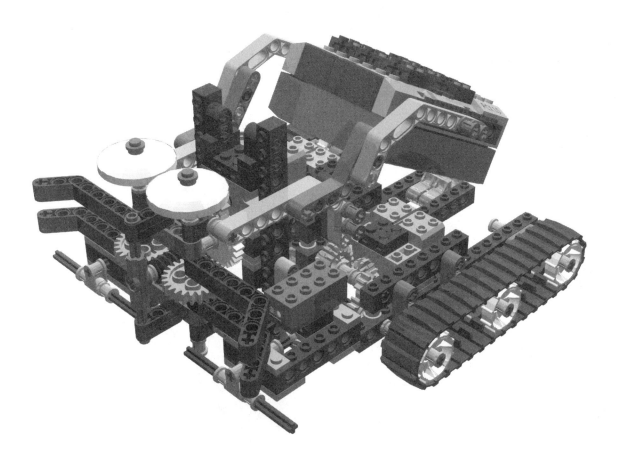

HunterBot is a treaded robot designed to locate and grab objects. It's based on one of my favorite robots–a wheeled robot with a claw, two RCX bricks, three touch sensors, two light sensors, and a rotation sensor. All of this hardware works towards the single objective of gathering pop cans.

Here's how it works. I wrapped a piece of blank, white paper around a few empty cola cans. I then placed the white cans in strategic spots on the floor. Using only the patterns on the floor to navigate, the robot would locate the cans, grab them, and check to see if its claws held something white. If the object wasn't a white can, the robot opened its claws and continued the search. Several schools have set up similar challenges in which the robots have to locate the cans and move them to a specified location. This is an excellent way to test your skills at building and programming with MINDSTORMS.

I designed HunterBot with this challenge in mind. The HunterBot's main features are the grabber arms, which can close firmly on objects, and the powerful tractor treads, which help the robot drag cans to another location. The bumpers play an important role, too. As HunterBot wanders around the floor, it will eventually bump into something–like a pop can, for example. When this happens, it turns and grabs the object.

SYNGRESS
syngress.com
A robot, of course, is only as good as its programming. Using the HunterBot program, which is an NQC program that you can download from the Syngress Solutions Web site (www.syngress.com/solutions), this robot can successfully grab and move cans or other small objects, should it happen to bump into them. Given an efficient search pattern, HunterBot should be able to bump into several cans. Consider this a starting point for a more sophisticated search strategy.

Your challenge is to observe HunterBot in action, think up ways to improve its ability to find cans, and modify the HunterBot program accordingly. Try altering the search pattern so that it's more efficient. If you fill the RIS Test Pad with objects, how long will it take HunterBot to remove them?

The Right Drive

The right drive sub-assembly includes the motor, gearbox, and right tractor tread. The gearbox reduces the rotation speed to a one–fifth of the motor's speed.

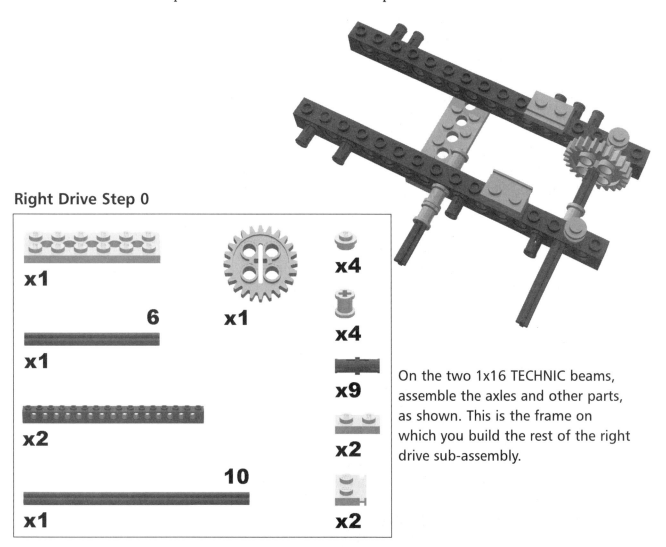

Right Drive Step 0

x1

6
x1

x2

10
x1

x1

x4

x4

x9

x2

x2

On the two 1x16 TECHNIC beams, assemble the axles and other parts, as shown. This is the frame on which you build the rest of the right drive sub-assembly.

Right Drive Step 1

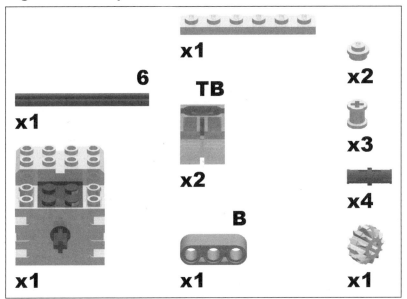

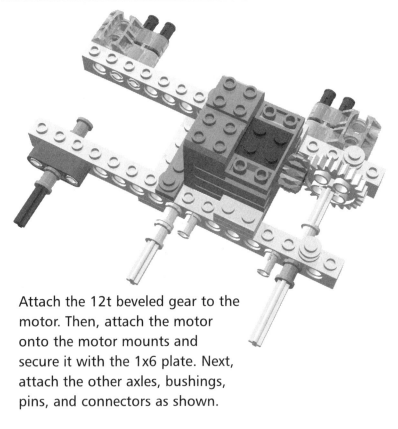

Attach the 12t beveled gear to the motor. Then, attach the motor onto the motor mounts and secure it with the 1x6 plate. Next, attach the other axles, bushings, pins, and connectors as shown.

Right Drive Step 2

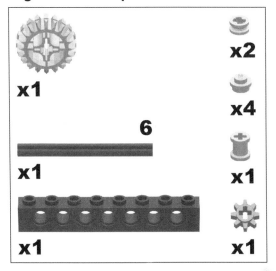

x1

6

x1

x1

x2

x4

x1

x1

Slide a #6 axle through the first hole of the 1x8 TECHNIC brick and secure the axle with a half-bushing. Then, slide the gears and bushings onto the axle as shown. The 8t gear should mesh with the 24t gear added in **Right Drive Step 0**, while the 20t beveled gear should mesh with the 12t beveled gear on the motor.

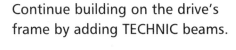

Continue building on the drive's frame by adding TECHNIC beams.

Right Drive Step 3

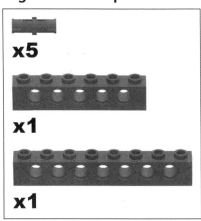

x5

x1

x1

Right Drive Step 4

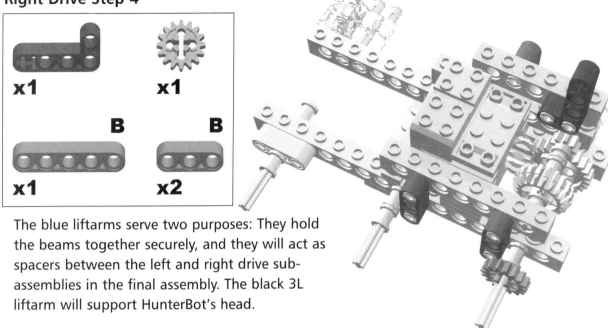

x1

x1

B **B**

x1 **x2**

The blue liftarms serve two purposes: They hold the beams together securely, and they will act as spacers between the left and right drive sub-assemblies in the final assembly. The black 3L liftarm will support HunterBot's head.

Right Drive Step 5

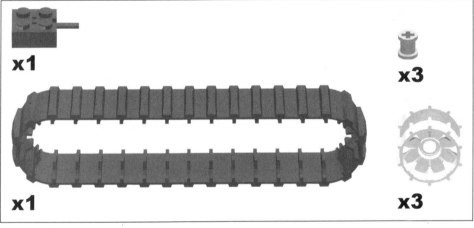

x1

x3

x1

x3

Add the sprocket wheels and the tractor tread as shown. Secure sprocket wheels with bushings. When you connect the wire to the motor, make sure that the wire points toward the back of the motor.

Now you can start the left drive sub-assembly.

Wheels or Treads?

HunterBot is designed to run on treads, but you could adapt it to run on wheels with very little modification.

The Left Drive

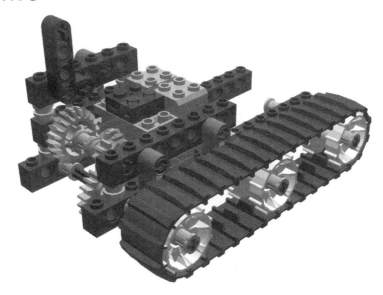

The left drive sub-assembly mirrors the right drive sub-assembly, but notice that the gears are exactly the same in both modules. This ensures that the wheels drive in the same direction when the motors run forward.

Left Drive Step 0

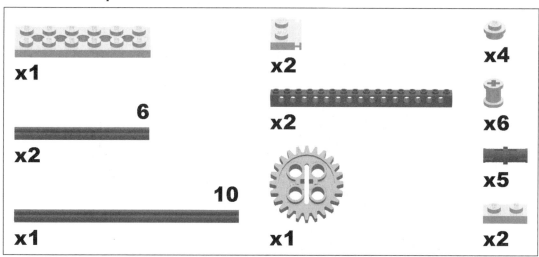

x1

6

x2

10

x1

x2

x2

x1

x4

x6

x5

x2

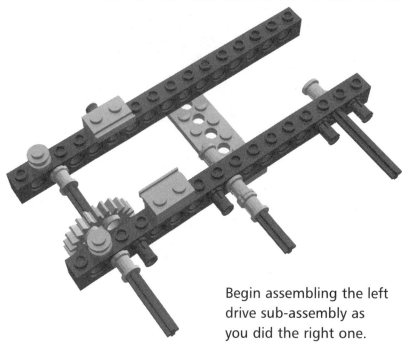

Begin assembling the left drive sub-assembly as you did the right one.

Left Drive Step 1

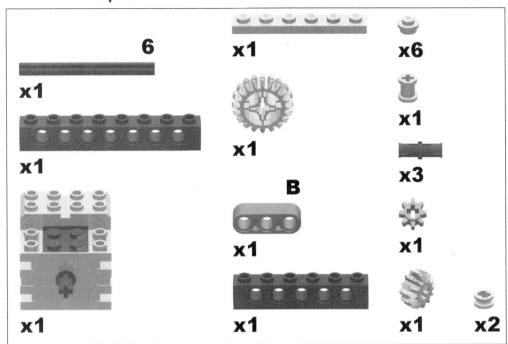

6

x1

x1

x1

x1

B

x1

x1

x6

x1

x3

x1

x1

x2

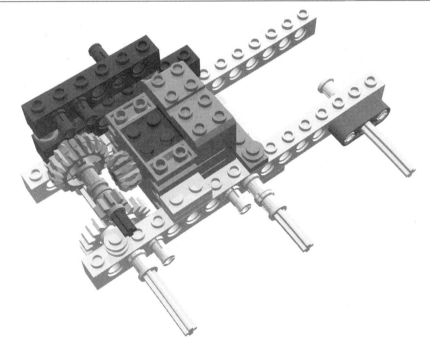

Left Drive Step 2

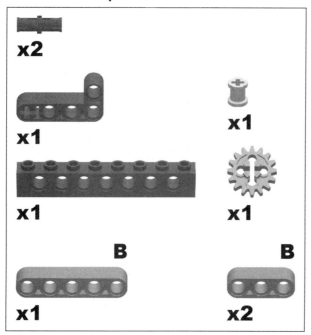

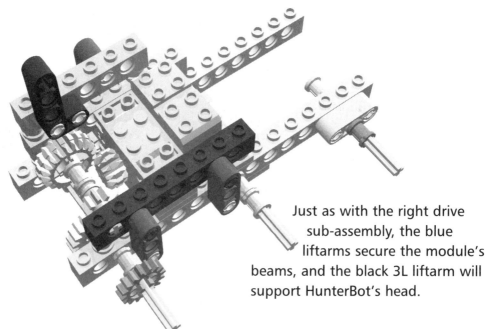

Just as with the right drive sub-assembly, the blue liftarms secure the module's beams, and the black 3L liftarm will support HunterBot's head.

Left Drive Step 3

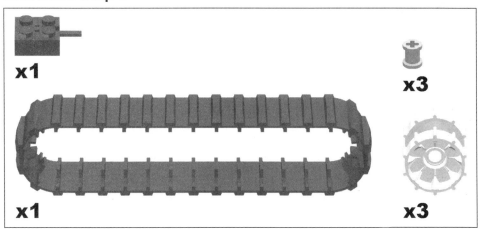

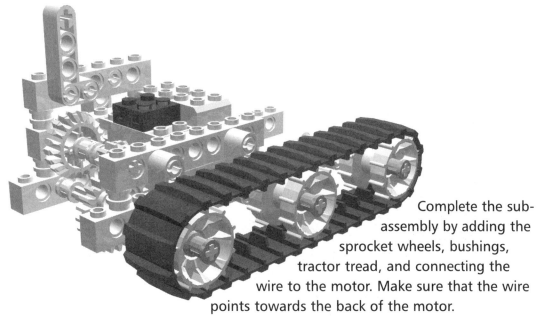

Complete the sub-assembly by adding the sprocket wheels, bushings, tractor tread, and connecting the wire to the motor. Make sure that the wire points towards the back of the motor.

Bricks & Chips...

Do You Have Clearance for That?

The LEGO treads look pretty cool, but they put your robot's body quite close to the ground. Consequently, HunterBot and other treaded robots may not have enough ground clearance to move easily on deep carpet.

The Grabber Arms

Now you begin the interesting part of HunterBot: the grabber arms. The business end of the grabber "claw" is a simple assembly of four angled liftarms on a series of axles. When the 24t gears rotate, the grabber arms open or close.

Inventing...

Arming Your Robot

For HunterBot's arms, I chose the black, angled liftarms. You may prefer the brightly-colored ones that came with the Ultimate Builders Set.

Grabber Arms Step 0

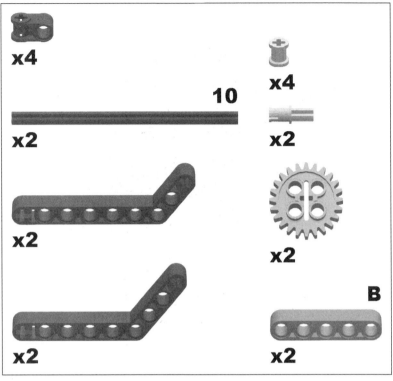

x4

10

x2

x2

x2

x4

x2

x2

B

x2

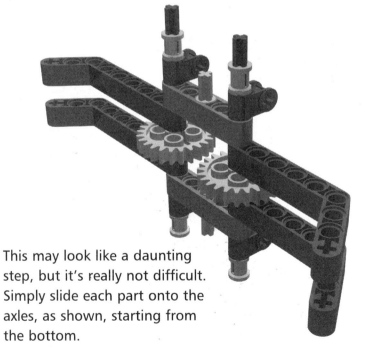

This may look like a daunting step, but it's really not difficult. Simply slide each part onto the axles, as shown, starting from the bottom.

Grabber Arms Step 1

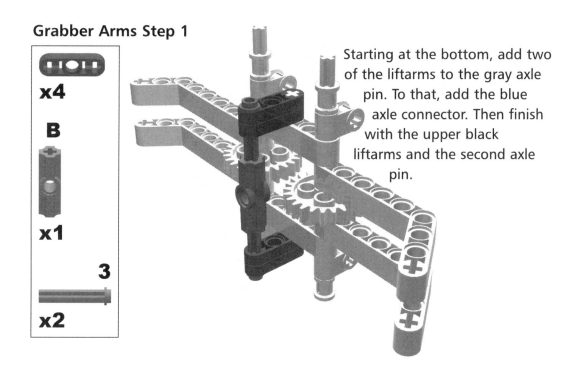

Starting at the bottom, add two of the liftarms to the gray axle pin. To that, add the blue axle connector. Then finish with the upper black liftarms and the second axle pin.

Grabber Arms Step 2

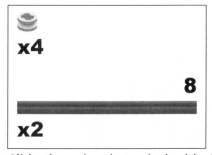

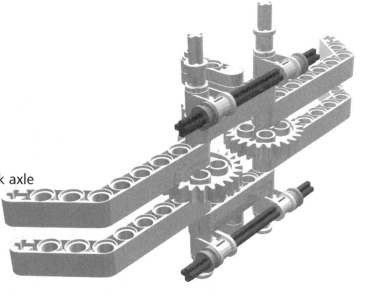

Slide the axles through the black axle joiners, then add the half-bushings. The axles provide the connection points for the grabber's motor sub-assembly, which you will build next.

The Grabber Motor

The grabber motor sub-assembly powers the grabber arms using a worm gear.

Grabber Motor Step 0

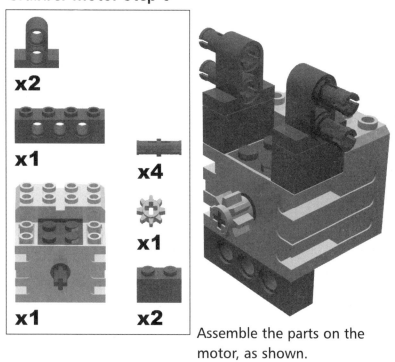

Assemble the parts on the motor, as shown.

Grabber Motor Step 1

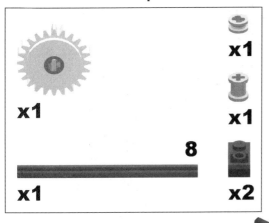

x1
x1
8
x1
x1
x1
x2

Slide the white clutch gear and the bushing onto the #8 axle, then attach it to the motor through the black 1x4 beam. A half-bushing holds the axle in place on the other side of the beam.

Bricks & Chips...

Using the Clutch Gear

A very important feature of this sub-assembly is the clutch gear. Without the clutch gear, the assembly simply wouldn't work—or worse, the motor could jam and become damaged. Make sure that you use the clutch gear and not a regular 24t gear.

Grabber Motor Step 2

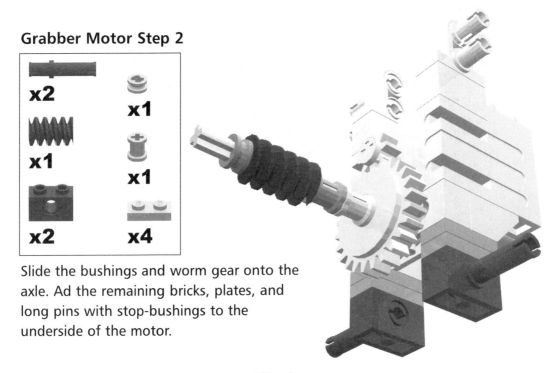

x2
x1
x1
x1
x2
x4

Slide the bushings and worm gear onto the axle. Ad the remaining bricks, plates, and long pins with stop-bushings to the underside of the motor.

Grabber Motor Step 3

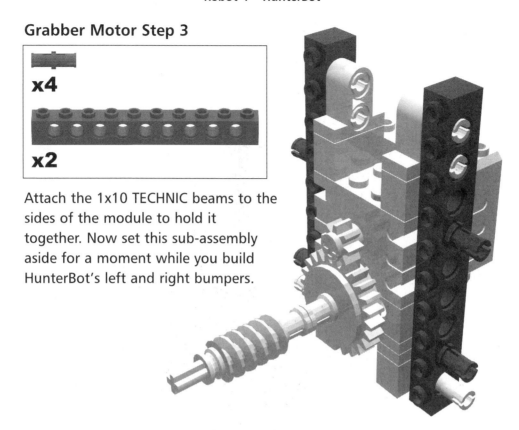

Attach the 1x10 TECHNIC beams to the sides of the module to hold it together. Now set this sub-assembly aside for a moment while you build HunterBot's left and right bumpers.

The Bumpers

These are probably the simplest bumpers you will ever build. While they aren't pretty, they work reliably, and they're very easy to put together. You will need to build two of these.

Bumper Step 0

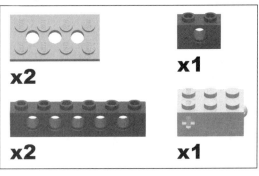

Attach the TECHNIC bricks and touch sensor to the two 2x4 plates with holes, as shown.

Bumper Step 1

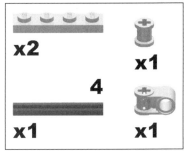

Slide the #4 axle through until it is almost touching the touch sensor's yellow button, and secure it with a bushing on the interior of the frame. Slide the gray axle joiner onto the front end of the axle, leaving enough space for the axle to slide in and out freely. If the axle can't push the button, your bumper won't work!

Bumper Step 2

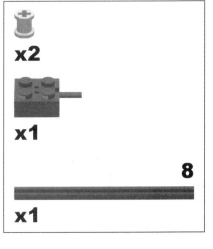

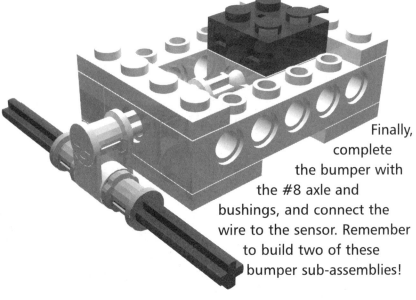

Finally, complete the bumper with the #8 axle and bushings, and connect the wire to the sensor. Remember to build two of these bumper sub-assemblies!

The RCX

The RCX brick needs a beam on either side so that it can be attached to the body of the robot.

RCX Step 0

x4

x1

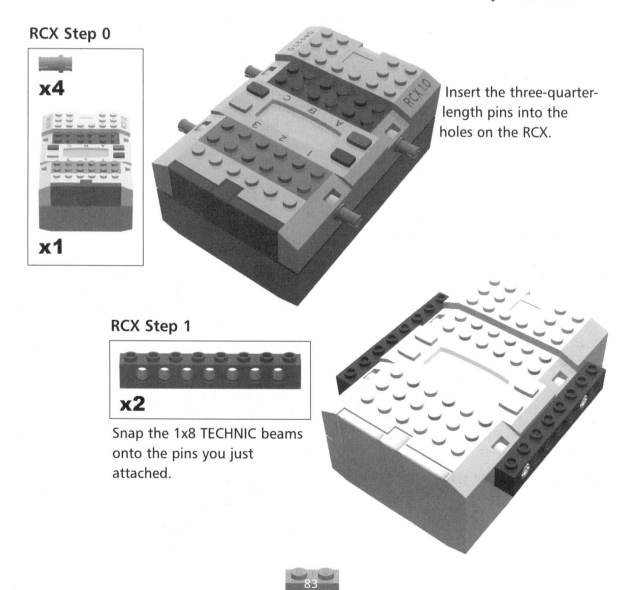

Insert the three-quarter-length pins into the holes on the RCX.

RCX Step 1

x2

Snap the 1x8 TECHNIC beams onto the pins you just attached.

Putting It All Together

This is the moment you've been working towards–the final assembly! This is the easiest yet the most satisfying part of building a robot.

First you'll create the HunterBot's head, which includes the grabber motor, grabber arms, and bumper sub-assemblies you built earlier, plus a light sensor. The various angled liftarms will connect the head to the body later in the final assembly.

Final Step 0

Locate the grabber arms sub-assembly that you built earlier.

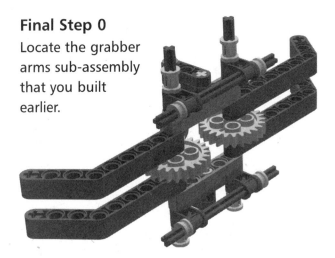

Final Step 1

This step may be a little tricky. Locate the grabber motor sub-assembly that you built earlier, and slip the motor's axle with the worm gear between the two 24t gears. The axle should fit securely inside the blue axle connector on the front of the grabbers. The worm screw should mesh with the gray gears.

As you do this, make sure that the left and right arms are open about the same distance.

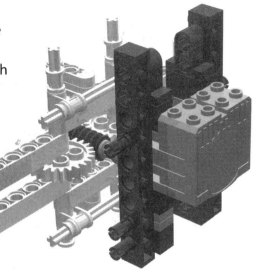

Final Step 2

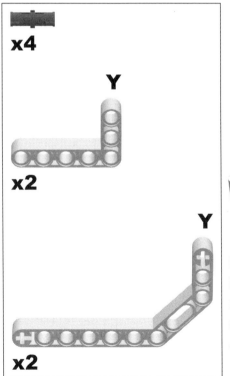

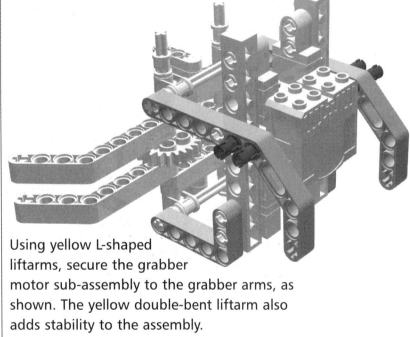

Using yellow L-shaped liftarms, secure the grabber motor sub-assembly to the grabber arms, as shown. The yellow double-bent liftarm also adds stability to the assembly.

Final Step 3

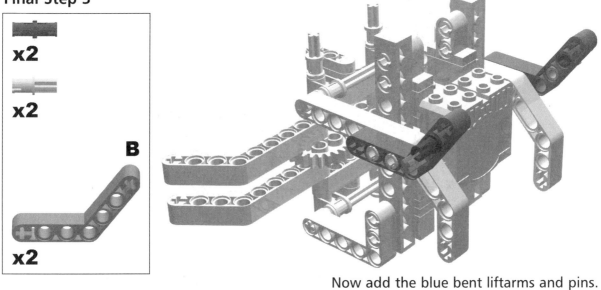

x2

x2

B

x2

Now add the blue bent liftarms and pins.

Final Step 4

Y   B

x2   x4

x2   x2

Attach two more of the double-bent yellow liftarms onto the blue liftarms, as shown.

Attach the eyes to the protruding axles of the grabber sub-assembly.

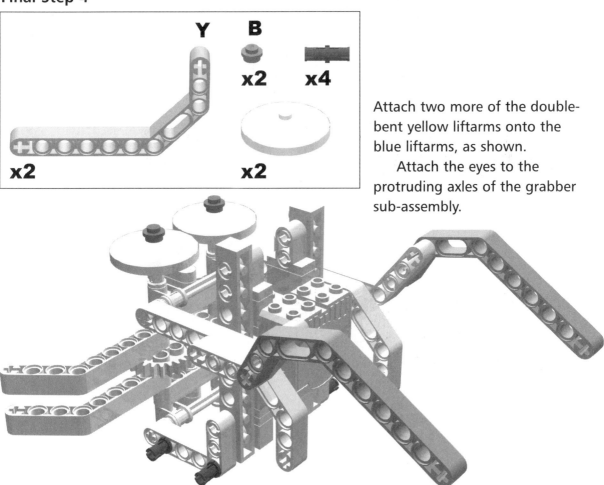

Final Step 5

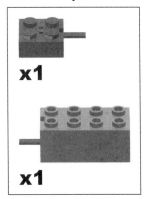

Finally, locate the two bumpers that you built earlier and snap them onto either side of the head. On the left bumper, attach the light sensor.

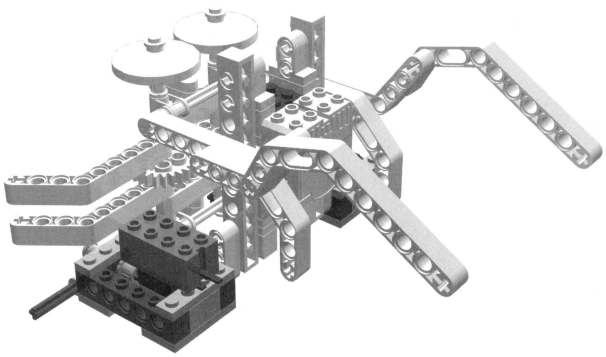

Developing & Deploying...
Using the Light Sensor

HunterBot uses the light sensor to prevent the arms from opening too far: The light sensor detects when the black grabber arm is near.

Final Step 6

Put the head you just built aside for a moment. Locate the right drive sub-assembly.

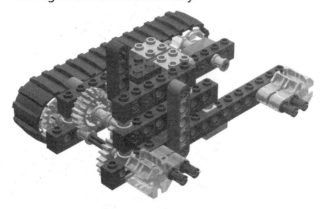

Final Step 7

Connect the two drive sub-assemblies together as shown.

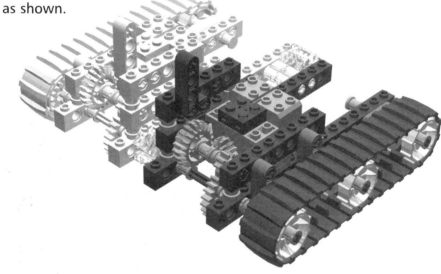

Inventing...

Space Available

Connector blocks, like the transparent blue ones that hold HunterBot's drive modules together, are often used to attach other accessories. Have a look at the RIS 2.0 *Constructopedia* for some ideas for things to attach to HunterBot.

Final Step 8

Position the head you finished in **Final Step 5** onto the black L-shaped liftarms, as shown. Secure it with four blue, long pins with stop-bushings.

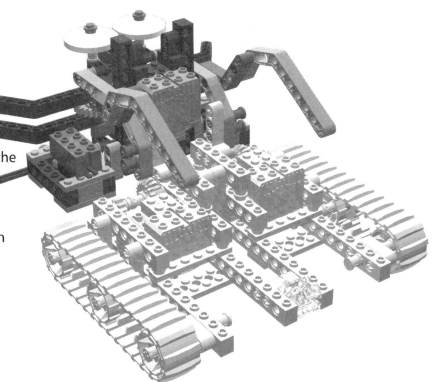

Final Step 9

Position the RCX sub-assembly as shown, and insert the four blue pins with stop-bushings to hold it in place.

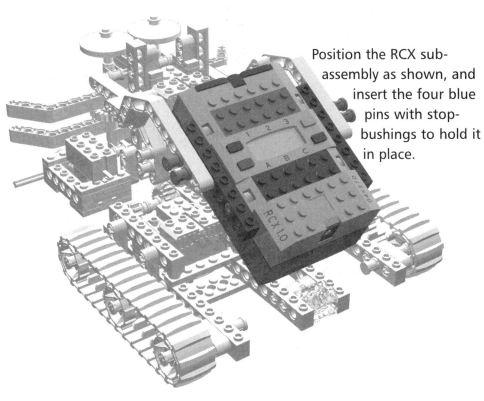

Bricks & Chips...

Using Long Pins with Stop Bushings

Long pins with stop bushings are excellent for attaching sub-assemblies in a way that makes it easy for you to remove them later. If you want to remove a sub-assembly—like the head, for example—simply pull out the pins.

Final Step 10

x6

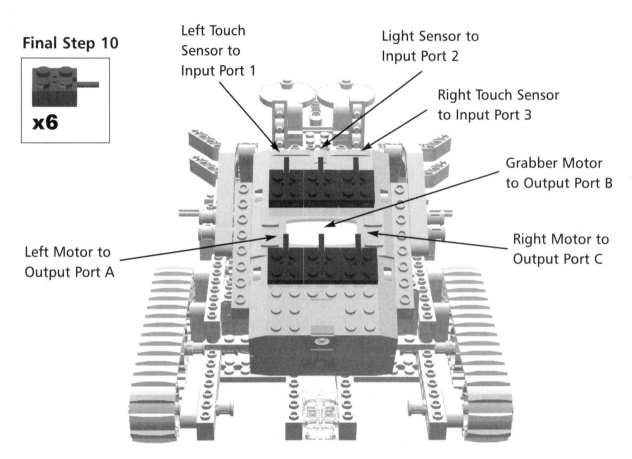

Left Touch Sensor to Input Port 1

Light Sensor to Input Port 2

Right Touch Sensor to Input Port 3

Grabber Motor to Output Port B

Right Motor to Output Port C

Left Motor to Output Port A

Connect the wires to the RCX brick as shown.

Connect the left and right touch sensors to Input Ports 1 and 3, and the light sensor to Input Port 2.

Connect the left motor to Output Port A and the right motor to Output Port C. Connect the grabber motor to Output Port B.

Make sure that the motor wires are oriented on the RCX as shown, with the wires pointing forward.

You're done! HunterBot can't wait to get out there and start grabbing things!

Developing & Deploying...

Programming HunterBot

You can download the NQC program for HunterBot from the Syngress Solutions Web site (www.syngress.com/solutions).

Robot 5

Nessie

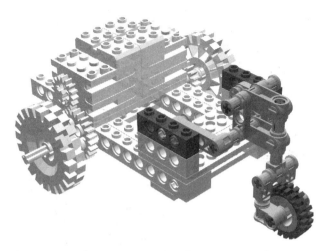

Nessie was built as a simpler yet versatile alternative to the RoverBot that can be found in the RIS 2.0 *Constructopedia*. Nessie is a small robot base that can be configured for a variety of functions. It makes a good beginner's model because it uses few parts and can be built within half an hour. In this implementation, Nessie is equipped with a dual light sensor for line following, for which it is eminently suitable. The two light sensors enable Nessie to negotiate a sharp 90-degree bend in both a clockwise and counter-clockwise direction. The two light sensors also create a variety of programming options.

Nessie is powered by two of the LEGO geared motors in a differential drive arrangement. The motors are placed in front, creating a sort of front-wheel drive, for better traction and ease of turning. A gear reduction of 3:2 (or 1.5:1) is used to give Nessie a balance between speed and reliability when used as a line follower. Nessie also features a trailing caster wheel, which gives it great maneuverability, which is an important factor for line following.

As we said, Nessie's base can be easily customized–we encourage you to try the following changes, observing the effect on the robot's performance:

- **Motor placement** Move the motors back relative to the driving wheels.

- **Gear ratio** Try different combinations of gears.

- **Wheels** Try different types of wheels.

- **Trailing wheel** Replace the trailing caster with other wheel arrangements, such as a sliding pulley wheel. We will show you how to build this type of wheel in Robot 6, Nellie. Nellie is a close cousin of Nessie.

- **Sensors** Attach bumpers that activate touch sensors, to turn the robot into an obstacle avoidance vehicle. We will also show you how to incorporate a rotation sensor when building Nellie.

The Caster Wheel

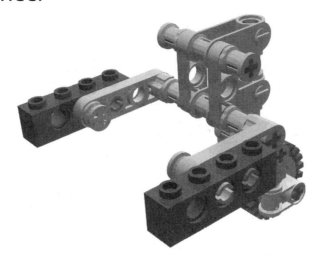

This is the caster wheel sub-assembly that gives Nessie its great maneuverability, which is so important for line following.

Bricks & Chips...

Building Caster Wheels

There are many ways to build a caster wheel. However, do not use a coupled caster wheel, as Mario Ferrari pointed out in his book *Building Robots with LEGO MINDSTORMS*. For this robot, it's best to use only a single, freely rotating wheel.

Caster Wheel Step 0

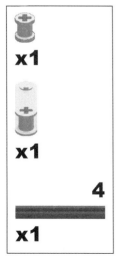

Start by building the caster wheel sub-assembly.

Caster Wheel Step 1

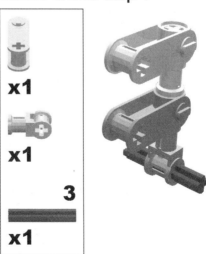

Caster Wheel Step 2

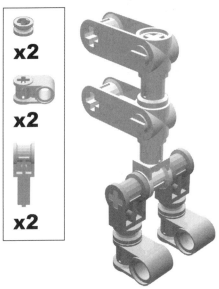

Caster Wheel Step 3

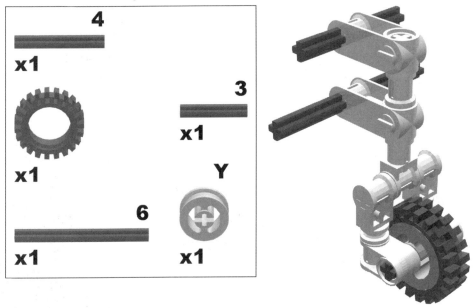

Caster Wheel Step 4

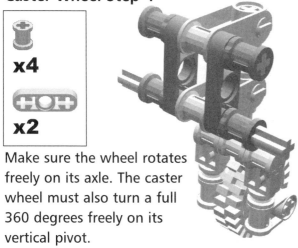

Make sure the wheel rotates freely on its axle. The caster wheel must also turn a full 360 degrees freely on its vertical pivot.

Caster Wheel Step 5

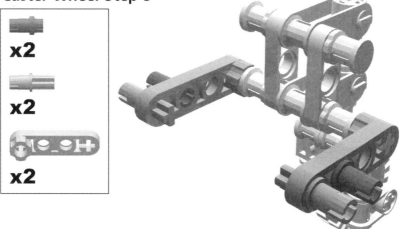

Caster Wheel Step 6

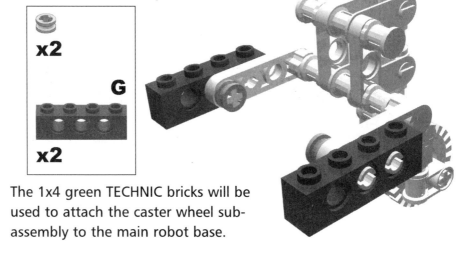

The 1x4 green TECHNIC bricks will be used to attach the caster wheel sub-assembly to the main robot base.

The Base

This is the main robot base, which can be customized for various functions.

Base Step 0

x4

x2

Start by building the side frames of the robot

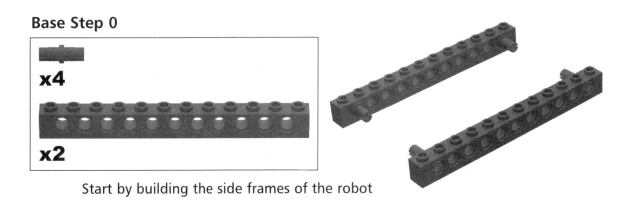

Base Step 1

x2

x2

8

x2

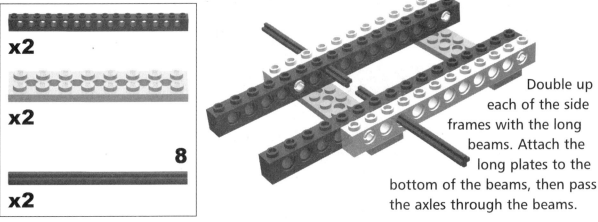

Double up each of the side frames with the long beams. Attach the long plates to the bottom of the beams, then pass the axles through the beams.

Base Step 2

x4

x2

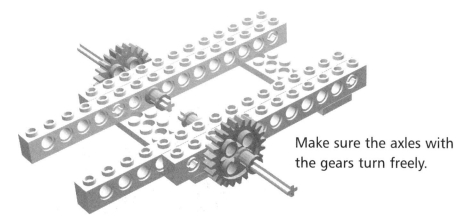

Make sure the axles with the gears turn freely.

Base Step 3

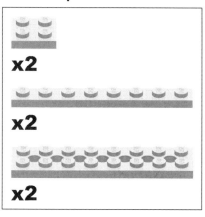

x2

x2

x2

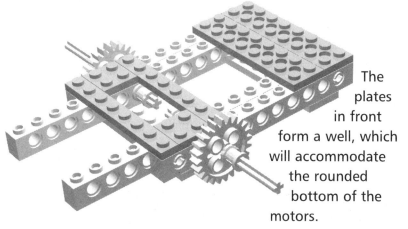

The plates in front form a well, which will accommodate the rounded bottom of the motors.

Base Step 4

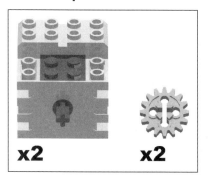

x2   **x2**

The motor has been raised one plate high to allow the 16t gear to mesh with the 24t gear.

Note that the gear meshing is not perfect, but it is close enough for our purposes.

Base Step 5

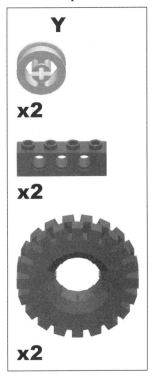

Y

x2

x2

x2

Bricks & Chips...

Using Wheel Variations

You may want to try other types of wheels, in particular the ones with the big wide tires.

Base Step 6

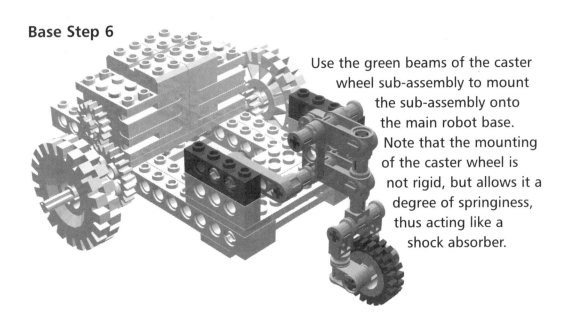

Use the green beams of the caster wheel sub-assembly to mount the sub-assembly onto the main robot base. Note that the mounting of the caster wheel is not rigid, but allows it a degree of springiness, thus acting like a shock absorber.

Base Step 7

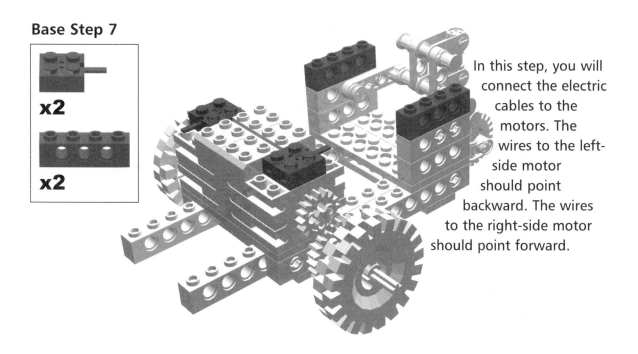

x2

x2

In this step, you will connect the electric cables to the motors. The wires to the left-side motor should point backward. The wires to the right-side motor should point forward.

Base Step 8

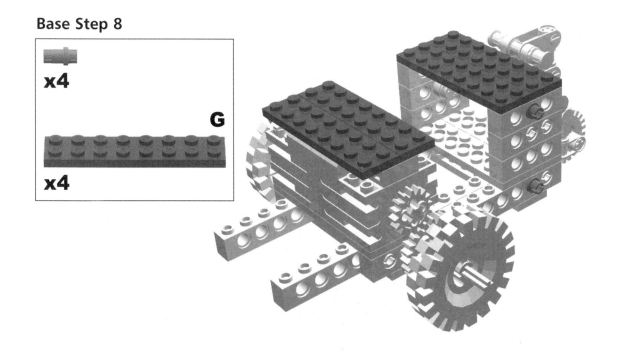

x4

G

x4

Base Step 9

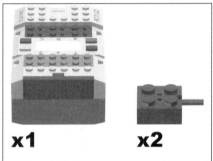

x2

Brace the back of the robot base in order to keep the caster wheel sub-assembly firmly in place.

Base Step 10

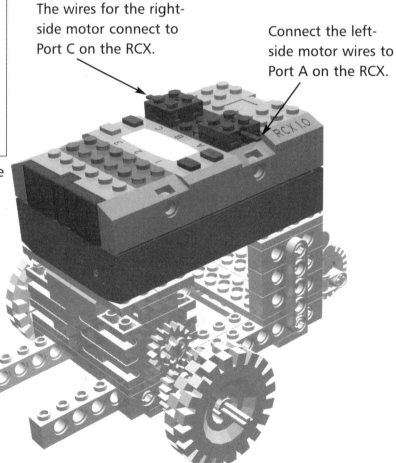

x1　　**x2**

The wires for the right-side motor connect to Port C on the RCX.

Connect the left-side motor wires to Port A on the RCX.

To attach the RCX, first separate the two sections. Put the top section aside. Place the bottom section on top of the green plates, and press down firmly. Replace the top section of the RCX (with the batteries), making sure the IR port faces forward.

Make sure the electric cables are oriented as shown.

Base Step 11

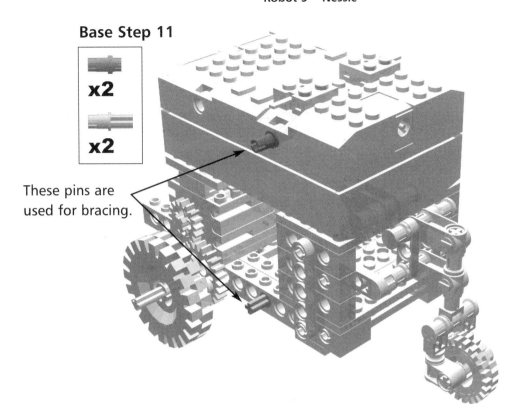

These pins are used for bracing.

Base Step 12

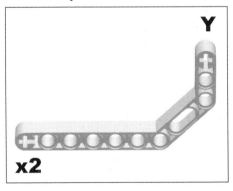

The yellow liftarm helps to keep the RCX in place. This completes the construction of the main robot base. You may want to add some decorative parts such as eyes or wings.

Bricks & Chips...

Testing Your Robot

Test the movements of Nessie using RCX built-in program #1. The robot should go forward at quite a fast speed. Next, test Nessie out using built-in program #4. Nessie should move in random directions, and should be able to turn freely on its caster wheel.

Adding the Light Sensors

With two light sensors, line following is a cinch. Nessie will negotiate a sharp 90-degree bend in both a clockwise and counter-clockwise direction. Furthermore, Nessie can be programmed to detect a T-junction on the line, where she can be made to stop, change direction, or carry out a specified task such as dumping a cargo into a receptacle.

Light Sensors Step 0

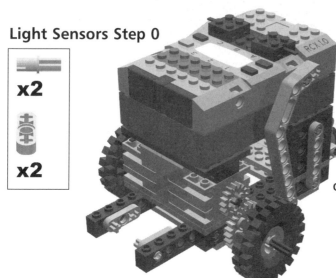

Start with the base sub-assembly.

Insert an axle pin into the front most hole of the beam extension. Place a 1x3 liftarm onto it. Do the same for the other side.

Light Sensors Step 1

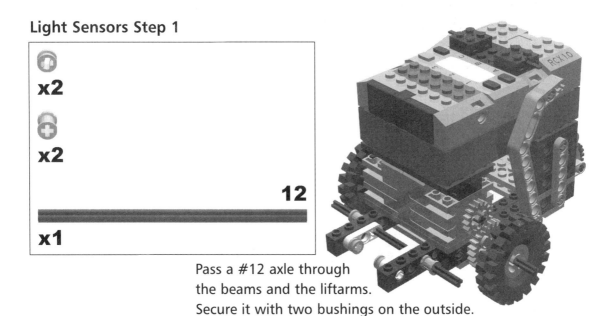

x2

x2

12

x1

Pass a #12 axle through
the beams and the liftarms.
Secure it with two bushings on the outside.

Light Sensors Step 2

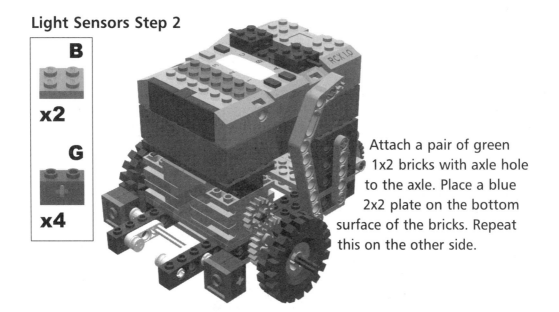

B

x2

G

x4

Attach a pair of green
1x2 bricks with axle hole
to the axle. Place a blue
2x2 plate on the bottom
surface of the bricks. Repeat
this on the other side.

Light Sensors Step 3

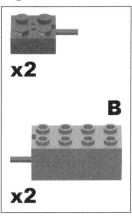

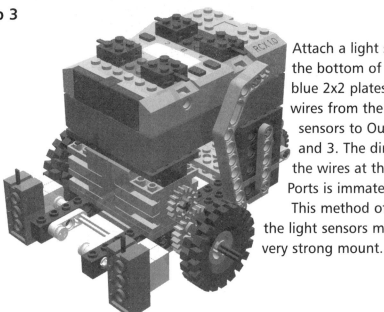

Attach a light sensor to the bottom of each of the blue 2x2 plates. Attach the wires from the light sensors to Output Ports 1 and 3. The direction of the wires at the Input Ports is immaterial.

This method of attaching the light sensors makes for a very strong mount.

Developing & Deploying...

Ready for More?

In Robot 6, Nellie, we will explore different building techniques to enhance line following. In Nellie, you will replace the caster wheel sub-assembly with a sliding wheel sub-assembly and add a rotation sensor. Not sure what this means? Have a friend build Nellie, and compare how Nessie and Nellie measure up in a line-following competition!

Robot 6

Nellie

At first glance, Nellie closely resembles Robot 5, Nessie. However, as you build this robot you will see that Nellie employs a significant change in how it moves by replacing the caster wheel with a sliding rear wheel. Nessie used a freely movable caster to give it great maneuverability for turning, which is advantageous for line following. However, as you may have already realized, following a straight line offers results that could be considered poor.

Nellie differs from Nessie by using a sliding rear wheel and an additional rotation sensor. The wheel rotates freely when the robot is going straight. The sliding wheel receives it name from the turning action of the robot. When Nellie makes a turn, the wheel doesn't rotate but instead slides (or skids) over the surface of the floor. Thus, for the sliding wheel to work there should be very little friction between the wheel and the floor. To achieve this minimal friction we will use a wheel without a tire. The addition of the rotation sensor, and a rear-sliding wheel will allow you to program your robot to travel a given distance. Nellie uses a rotation sensor paired with a light sensor, and can be programmed to follow a line until it reaches a designated point where it can then carry out a specified task.

The Slider

The sliding wheel and rotation sensor are incorporated in this sub-assembly, which is then attached to the main robot base.

Slider Step 0

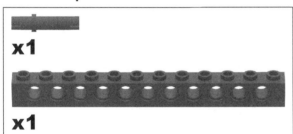

x1

x1

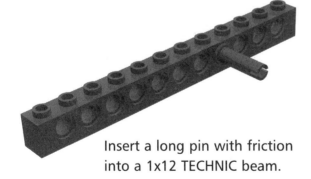

Insert a long pin with friction into a 1x12 TECHNIC beam.

Slider Step 1

B

x1   **x1**

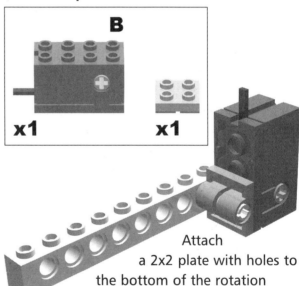

Attach a 2x2 plate with holes to the bottom of the rotation sensor. Be sure to position the rotation sensor with the wires pointing upward.

Push the plate onto the free end of the long pin with friction. Align the axle hole of the rotation sensor with the holes in the TECHNIC beam.

Slider Step 2

x1

8

x1

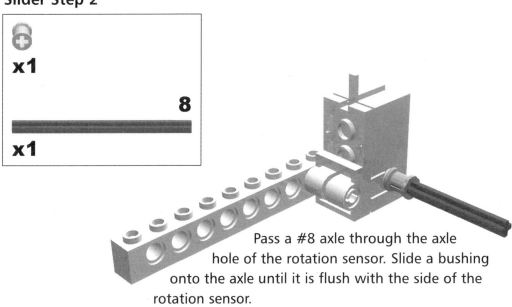

Pass a #8 axle through the axle hole of the rotation sensor. Slide a bushing onto the axle until it is flush with the side of the rotation sensor.

Slider Step 3

x1

x1

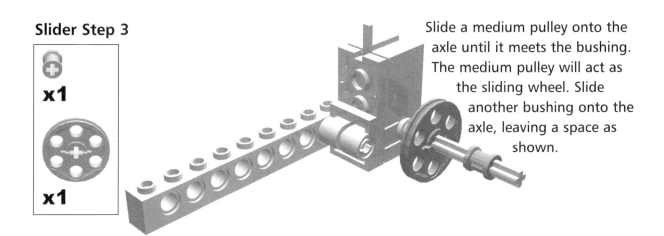

Slide a medium pulley onto the axle until it meets the bushing. The medium pulley will act as the sliding wheel. Slide another bushing onto the axle, leaving a space as shown.

Slider Step 4

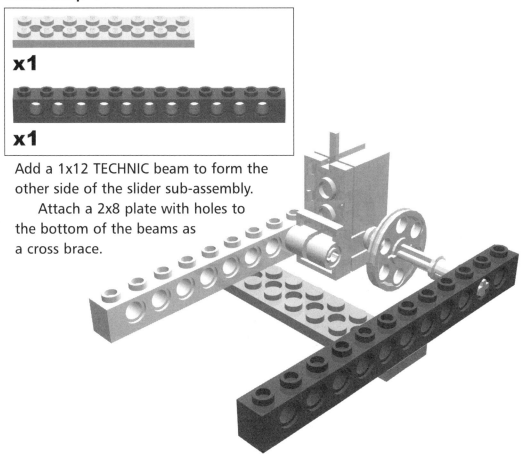

x1

x1

Add a 1x12 TECHNIC beam to form the other side of the slider sub-assembly.

Attach a 2x8 plate with holes to the bottom of the beams as a cross brace.

Developing & Deploying…

Increasing the Reliability of the Rotation Sensor

For the rotation sensor to track accurately, the wheel must stay in contact with the floor at all times. This requires a wheel with a tire. However, adding a tire will degrade the turning ability of the robot.

There is a trade off between maneuverability and accuracy of the rotation sensor tracking.

Normally, the rotation sensor will track properly with a plain wedge belt wheel (or medium pulley) as the sliding wheel. To increase the reliability of rotation sensor tracking, a white band can be fixed around the groove of the wedge belt wheel. If slipping of the wheel is a problem, then an O-ring tire should be fitted to the wedge belt wheel.

Try using Nellie with and without the white band as a tire on the pulley. See which assembly suits your needs best.

The Base

This is the main robot base of Nellie. This is similar to the base that was built in Robot 5, Nessie. Be sure to follow these steps closely as there are minor variations between the two.

Base Step 0

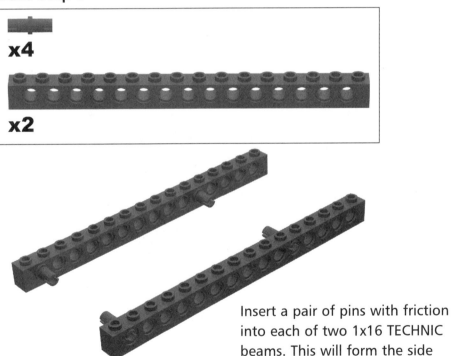

Insert a pair of pins with friction into each of two 1x16 TECHNIC beams. This will form the side frames of the robot base.

Base Step 1

x2

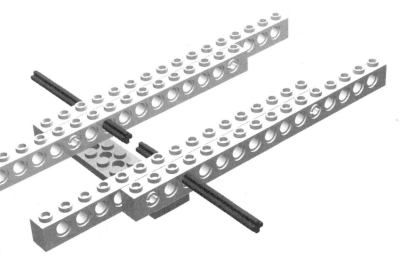

Snap another pair of 1x16 TECHNIC beams to the frame. Doubling the beams like this allows for better purchase for the driving axles.

The beam extension at the front will be used for mounting the light sensors. The extension at the rear is to provide clearance for the slider sub-assembly.

Base Step 2

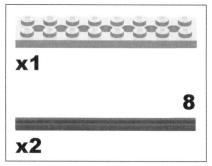

x1

8

x2

Attach a 2x8 plate with holes to the bottom of the beams.

Pass the #8 axles through each side of the frame. This will provide the cross bracing for the frame.

Base Step 3

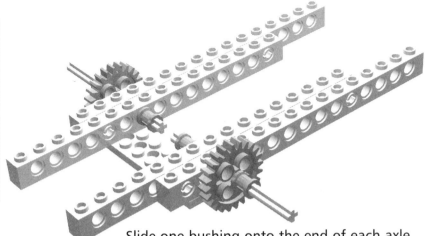

Slide one bushing onto the end of each axle located in the interior of the base frame.

Then, moving to the ends of the axles on the outside of the frame, slide a 24t gear onto each of the axles and secure the gears with the remaining bushings.

Base Step 4

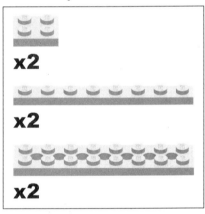

Attach two 2x8 plates with holes four studs from the end of the rear TECHNIC beams as shown.

Next, attach the two 1x8 plates and two 2x2 plates to form the support frame for the motors.

Base Step 5

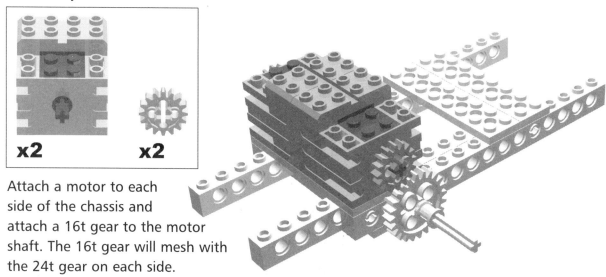

Attach a motor to each side of the chassis and attach a 16t gear to the motor shaft. The 16t gear will mesh with the 24t gear on each side.

The meshing of the gears is not exact LEGO geometry; however, for practical purposes it is close enough.

Base Step 6

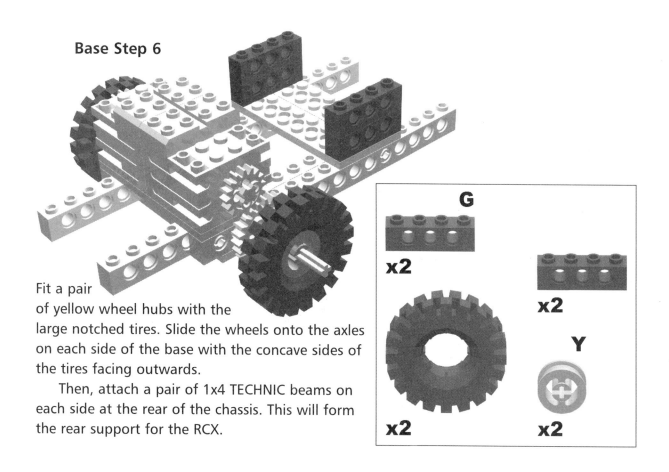

Fit a pair of yellow wheel hubs with the large notched tires. Slide the wheels onto the axles on each side of the base with the concave sides of the tires facing outwards.

Then, attach a pair of 1x4 TECHNIC beams on each side at the rear of the chassis. This will form the rear support for the RCX.

Base Step 7

Attach the slider sub-assembly to the base. Check that the sliding wheel is able to rotate freely.

Base Step 8

x2

x2

Add another 1x4 TECHNIC beam to each of the stacks of 1x4 beams attached in **Base Step 6**. Attach a short connector wire to each of the motors.

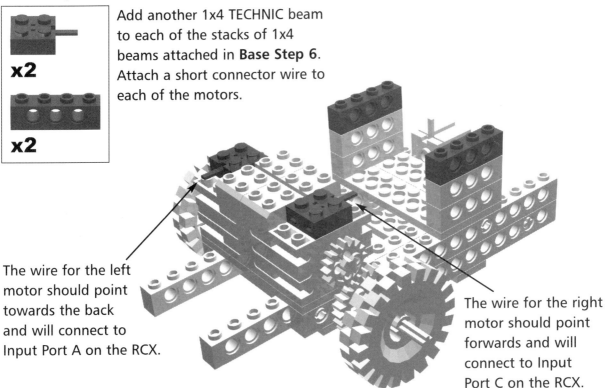

The wire for the left motor should point towards the back and will connect to Input Port A on the RCX.

The wire for the right motor should point forwards and will connect to Input Port C on the RCX.

Base Step 9

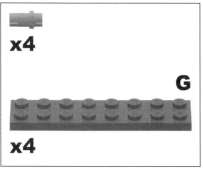

x4

G

x4

Place four 2x8 green plates to form the support for the RCX. Insert a pair of dark gray three-quarter length pins to each side of the vertically stacked 1x4 beams.

Base Step 10

x2

Attach a 1x5 liftarm on each side of the stack of 1x4 beams. These liftarms provide vertical bracing for the base.

Base Step 11

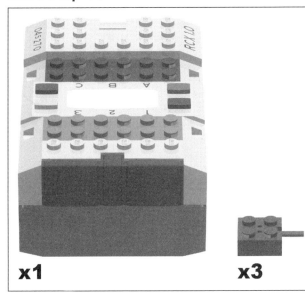

x1   **x3**

Attach the RCX to the top supporting plates of the robot base. To do so, separate the two sections of the RCX. Position the bottom section over the supporting plates and press down firmly. Replace the top section of the RCX (with the batteries) carefully over the bottom section.

Connect the free ends of the electric cables from the left motor to Input Port A, and attach the wire for the right motor to Input Port C.

The wire from the rotation sensor is connected to Input Port 2 on the RCX. The direction of this wire is immaterial.

Bricks & Chips...

Mounting the RCX

It is not good practice to simply press the whole body of the RCX onto the chassis mount as it may damage the insides.

Base Step 12

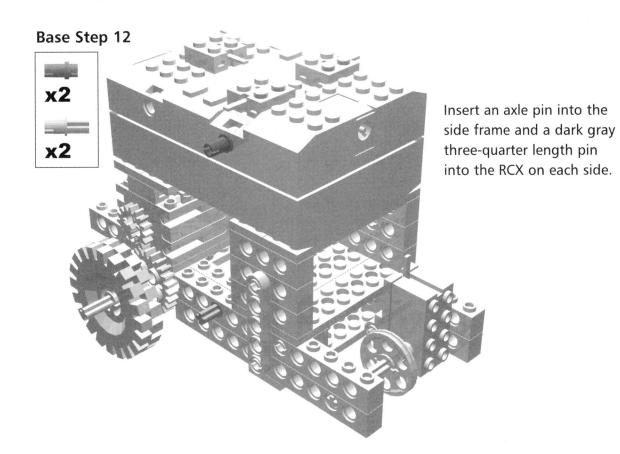

Insert an axle pin into the side frame and a dark gray three-quarter length pin into the RCX on each side.

Base Step 13

Y

x2

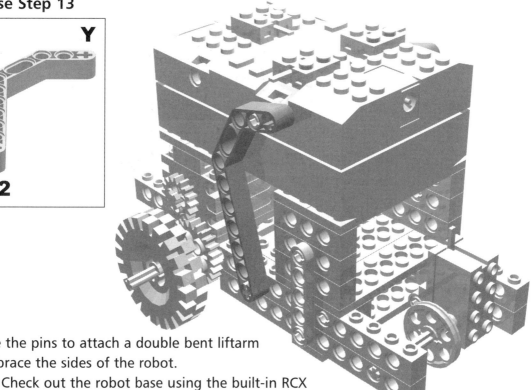

Use the pins to attach a double bent liftarm to brace the sides of the robot.

Check out the robot base using the built-in RCX Program #1. Nellie should move forward in a reasonably straight line. You can also use the built-in RCX Program #4 to see how the robot turns with the sliding wheel.

The robot can be used as it is, or you can attach a light sensor or even two light sensors. We show how to attach dual light sensors in the next sub-assembly. With two light sensors and the rotation sensor, this robot can be used in competitive challenges.

Bricks & Chips...
Benefits of Bracing

It is always good practice to brace your robot as much as possible, so that if you pick up your robot by holding the RCX, the rest of the robot will not come off.

The Light Sensors

The dual light sensors are identical to those found on Nessie.

Light Sensors Step 0

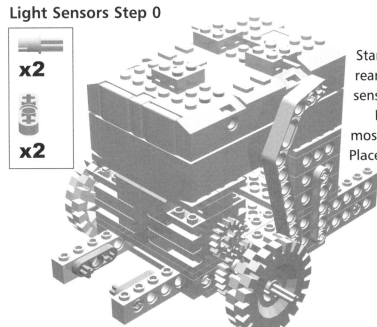

Start with the main robot base with rear sliding wheel and rotation sensor.

Insert an axle pin into the front most hole of the beam extension. Place a 1x3 liftarm onto it. Do the same for the other side.

Light Sensors Step 1

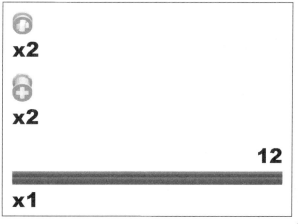

x2

x2

12

x1

Pass a #12 axle through the beams and the liftarms. Secure it with two bushings on the outside.

Light Sensors Step 2

B

x2

G

x4

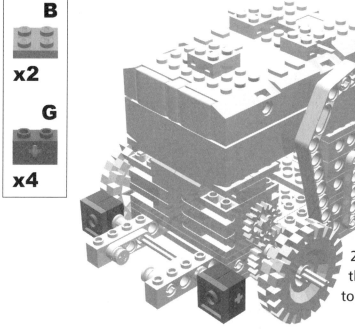

Fix a pair of green 1x2 bricks with axle holes to the free end of the axle. Place a blue 2x2 plate on the bottom surface of the 1x2 bricks to hold them together. Repeat for the other side.

Light Sensors Step 3

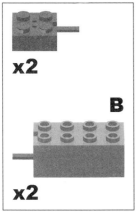

x2

B

x2

Attach a light sensor
to the bottom of
each of the blue 2x2
plates. For some reason, the light
sensors are more stable when they
are attached with the stud surface of
the sensor adjoining the bottom surface
of another brick.

Attach the wires from the light sensors to Input
Ports 1 and 3 respectively. The direction of the wires
at the input ports is immaterial.

This method of attaching the light sensors gives
a very strong mount

Robot 7

The DominoBot

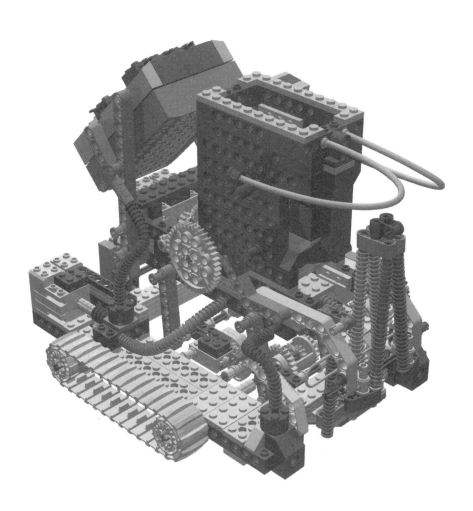

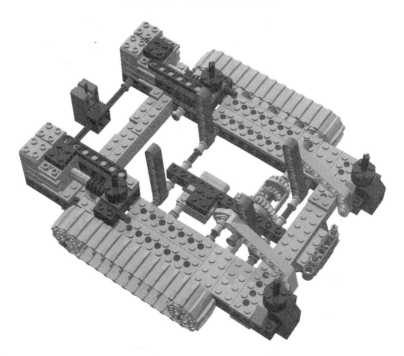

The process of laying dominoes by hand takes a great deal of time, accuracy, and patience. I was interested in seeing what it would take for a robot to do the same. The idea behind DominoBot was to create a robot that could be loaded with dominoes and program it to follow a black line on a white background while laying dominos at a preset distance along the path. This proved to be more of a challenge than I imagined!

In order to build this robot, you will require a set of dominos. The dominoes used for this design are Double Nine Color Dot Dominoes. These dominoes are approximately four LEGO studs in width by seven studs in height. It is possible that other dominoes of similar size will work as well.

To perform this task, some important details had to be worked out. There are three major components to the DominoBot that enable it to accomplish its many simultaneous actions. First, there is the *drive mechanism*, which consists of left and right track drive units that serve the purpose of moving the DominoBot in any direction. The drive also contains the second major component, the *distance sensor*. The distance sensor uses a differential connected between the two drive units. This component detects the distance the DominoBot has traveled by way of registering clicks to a touch sensor (set up as a *PULSE* touch sensor in NQC). The differential allows for clicks to be registered only when both drive tracks are moving forward; it does not register clicks if the DominoBot is turning. This was an important feature since I did not want the dominos being placed too closely or too far apart.

The third major components are the *loader* and *loader arm*. These parts are responsible for the main domino-placement action. The loader takes a stack of dominoes and places each one. It does this by use of a touch sensor to determine when a domino has been ejected out of the loader and also when the loader arm has reset itself. The loader arm proved to be the greatest challenge in designing this robot. Since the dominoes did not fit snugly in any part of the loader mechanism, I had to find a way to accurately and carefully place each domino as it slid out of the *loader.* Some patient experimenting revealed that the LEGO flexible hoses would do the trick for the placement process. The loader arm consists mostly of these flexible hoses arranged in such a way that they form a path for the domino to slide down and stand up on the ground. The ability of the hoses to flex allows for the orientation of each domino to be corrected if it does not come out of the loader smoothly.

Another challenge was the limited number of available motors for this project. I had to find a way to make all this work with one motor (the two others are used to drive the robot). The design that I came up with allows for all domino-placement actions to be done with one motor and touch sensor. The loader and loader arm both run in sync with each other from the motor, while the touch sensor is used to inform the RCX when a domino has been released and when the loader mechanism has reset itself.

The NQC program for the DominoBot is available for your use and modification on the Syngress Solutions Web site (www.syngress.com/solutions). Programming DominoBot consists of a few core routines. When DominoBot first begins, its will move left to right in a sweeping patter while it calibrates light sensor values. Since the LEGO light sensor reads light values based on surrounding light, it is hard to use specific values to identify black and white.

Once it has calibrated light, it begins moving forward along the black line. This action is monitored by the *checklight()* task. While doing this the distance sensor is incremented each time it is tripped by the cam gear. Testing revealed that three clicks would be an adequate distance for placing each domino. The *checkdistance()* task is responsible for this.

Once the *checkdistance()* task counts three clicks, it calls on the *dodomino()* task. This task stops all other tasks while it proceeds to place a domino. It makes sure that the loader arm has reset itself so that it does not clip the just-placed domino when beginning to move forward again.

All the above tasks are repeated for each domino in the loader. You will have to keep the loader full of dominos while DominoBot is navigating the course. You will also have to stop DominoBot once the dominos have all been placed.

The Loader

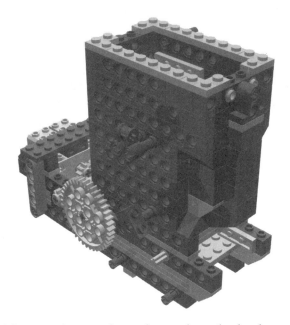

The loader sub-assembly delivers a domino from the stack to the loader arm by using a push mechanism. It also uses a touch sensor to alert the program that a domino has been delivered.

Loader Step 0

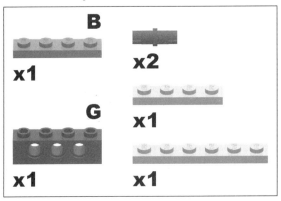

Take note of these two friction pins. In **Final Step 1**, these will be the joining point for the two flexible hoses used to deliver each domino to the loader arm.

Loader Step 1

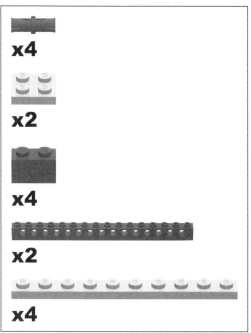

x4

x2

x4

x2

x4

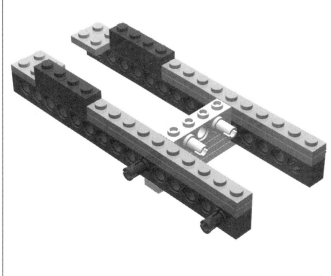

Loader Step 2

x1

x1

x2

x1

x2

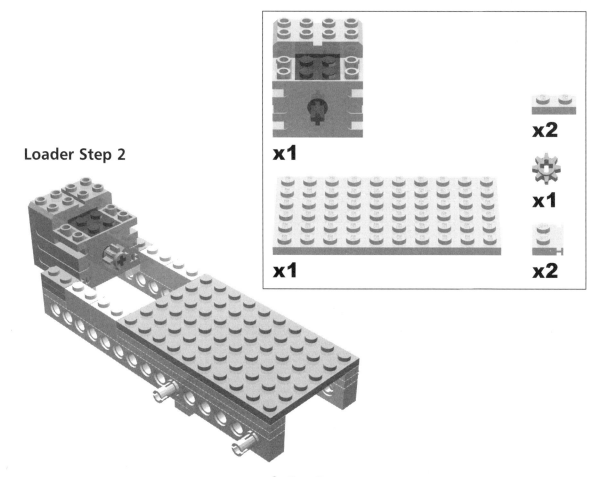

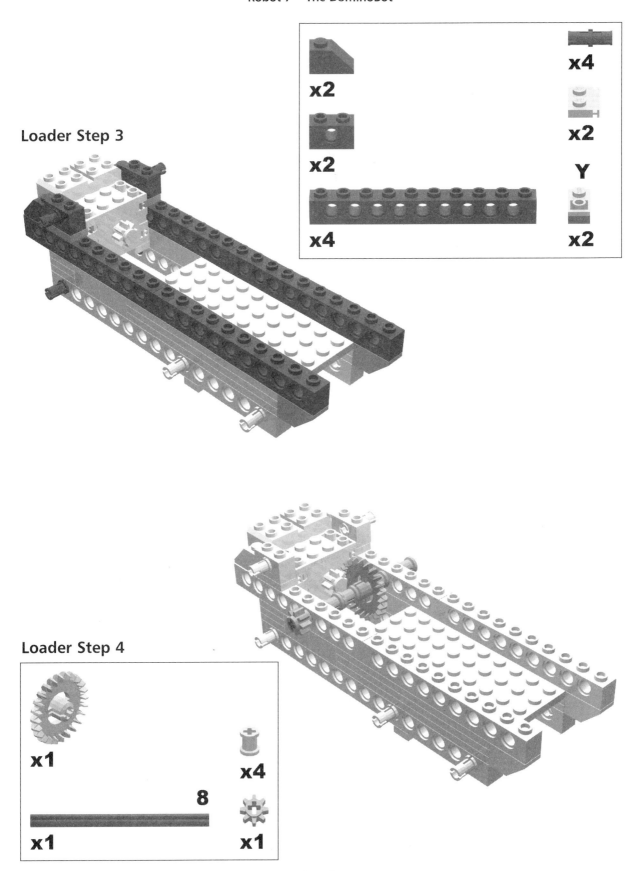

Loader Step 3

Loader Step 4

Loader Step 5

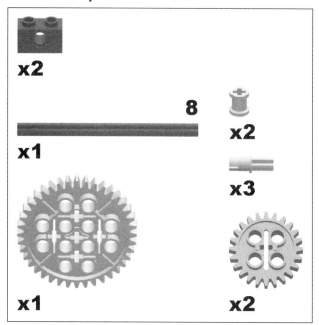

8

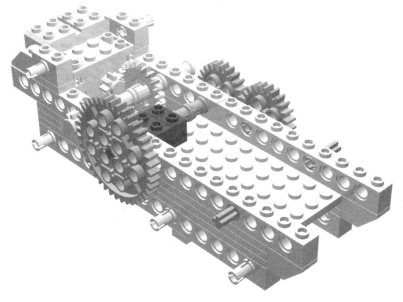

Loader Step 6

Y

x1

x1

x2

x1

The 1x3 plate that you insert into the inner wall helps keep the domino somewhat straight in the loader and prevents it from being dispensed at an angle.

Loader Step 7

x2

x2

x1

G

x1

x1

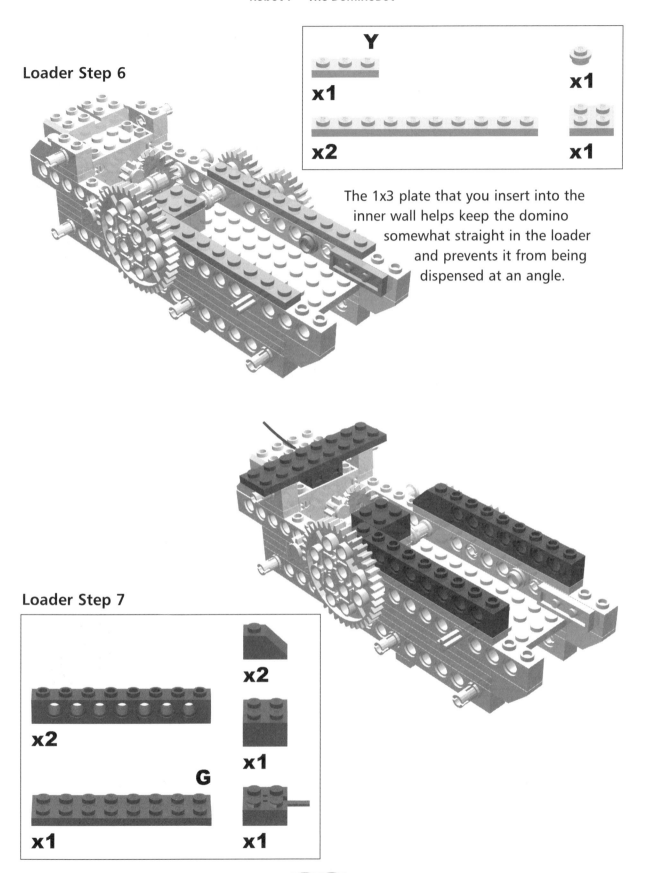

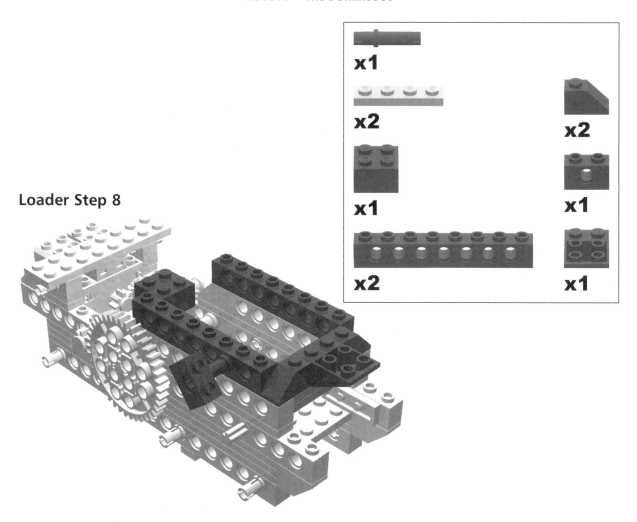

Loader Step 8

Loader Step 9

Y

x2

x2

x2

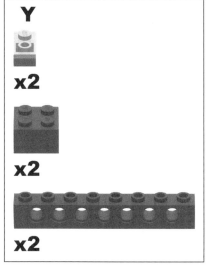

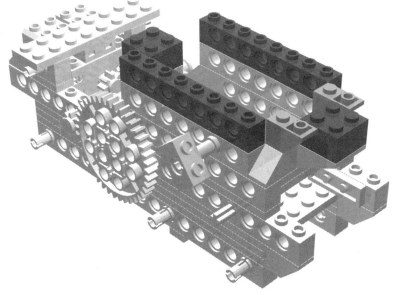

Loader Step 10

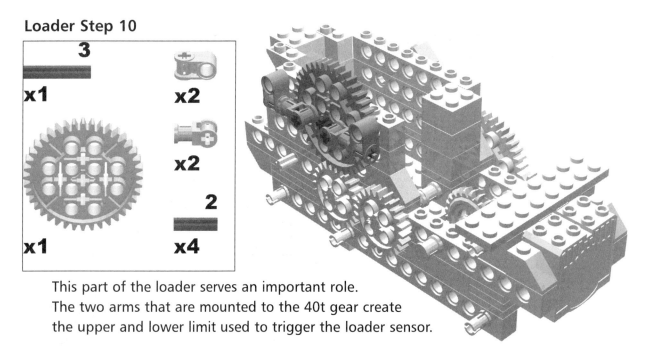

This part of the loader serves an important role.
The two arms that are mounted to the 40t gear create
the upper and lower limit used to trigger the loader sensor.

Loader Step 11

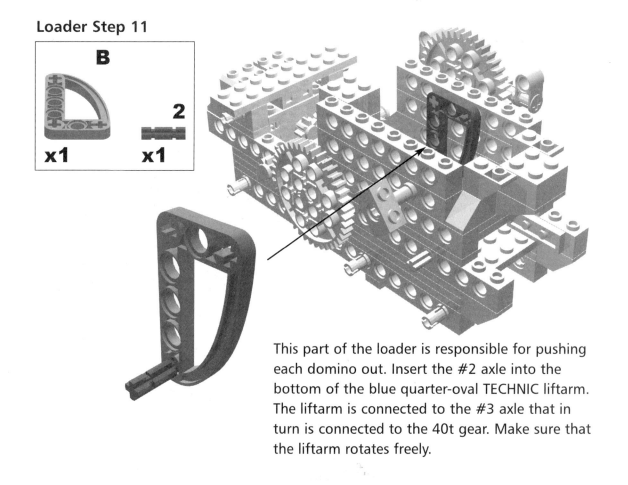

This part of the loader is responsible for pushing
each domino out. Insert the #2 axle into the
bottom of the blue quarter-oval TECHNIC liftarm.
The liftarm is connected to the #3 axle that in
turn is connected to the 40t gear. Make sure that
the liftarm rotates freely.

Bricks & Chips..

Loader Sensors & Limit..

...sensor is all that is needed to detect ...ed, as well as when the loader has ...e the next domino.

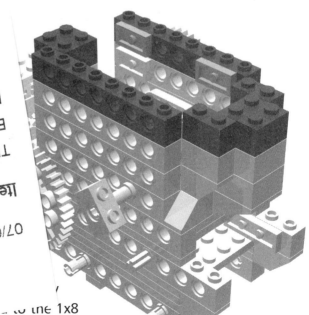

x2

x2

x2

x2

...to the 1x8 ...TECHNIC beam.

Terminal # 26

Join Summer Reading: www.saclibrary.org/src

DUE DATE 07-24-15
BARCODE 31652002176676
TITLE 10 cool Lego Mindstorms

DUE DATE 07-24-15
BARCODE 33029072597917
TITLE The LEGO Mindstorms

DUE DATE 07-24-15
BARCODE 33029070766673
TITLE Basic robot building : with

DUE DATE 07-24-15
BARCODE 33029099170771
TITLE Getting to know Lego

DUE DATE 07-24-15
BARCODE 33029048222087
TITLE 10 cool Lego Mindstorms

Item(s) Checked Out

07/03/2015

Loader Step 14

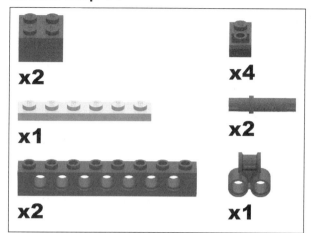

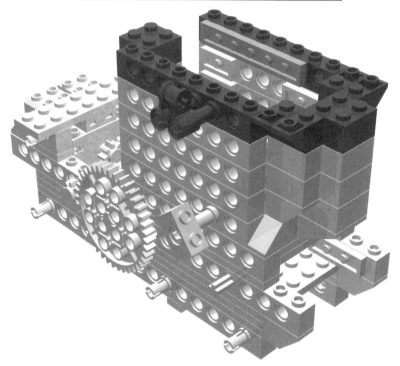

Loader Step 15

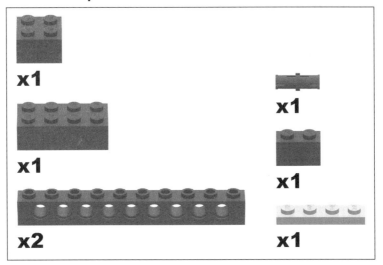

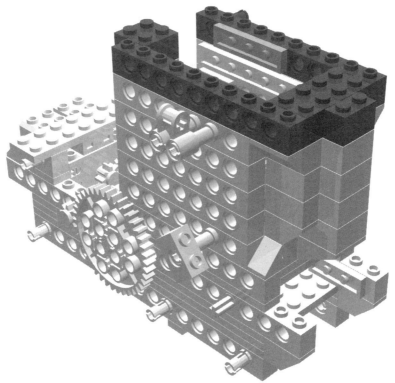

Loader Step 16

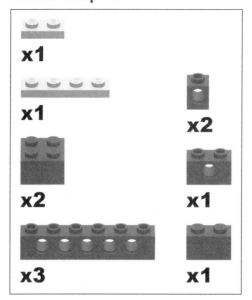

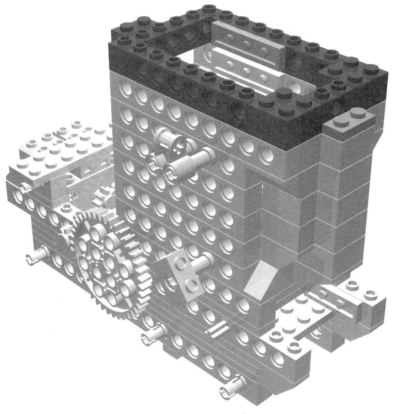

Loader Step 17

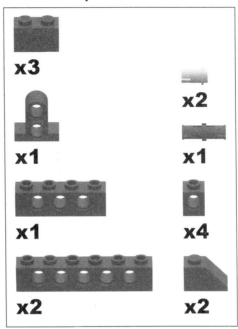

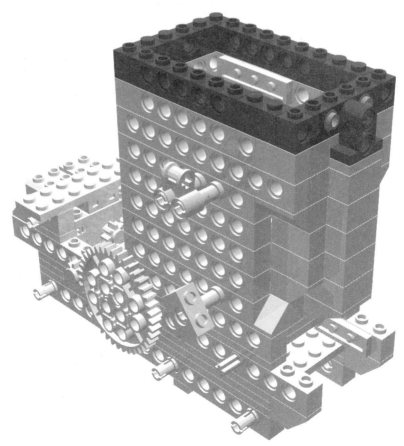

Loader Step 18

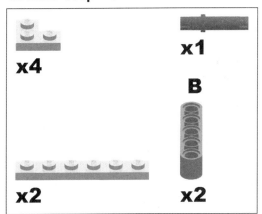

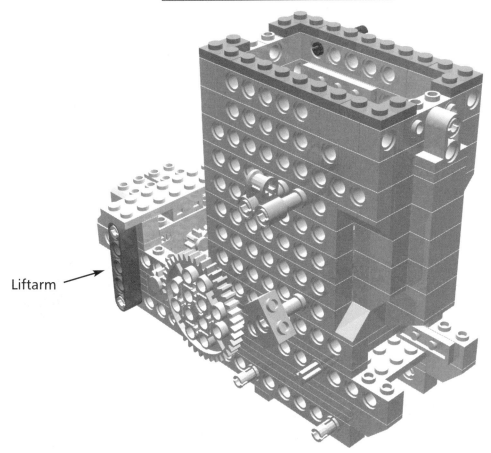

Liftarm

The Right Drive

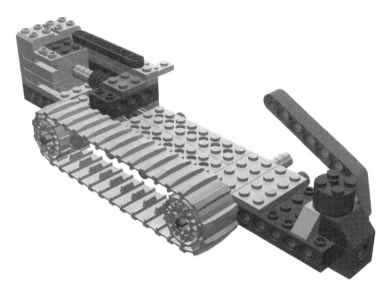

The right drive sub-assembly provides motion to the DominoBot. Tracks provide the ability to turn on a tight radius.

Right Drive Step 0

x2

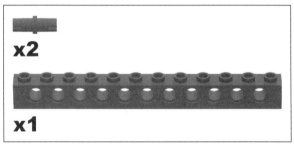

x1

Right Drive Step 1

x1

Right Drive Step 2

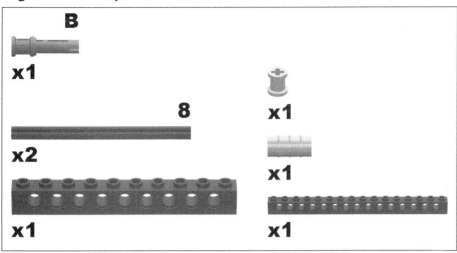

B

x1

8

x2

x1

x1

x1

x1

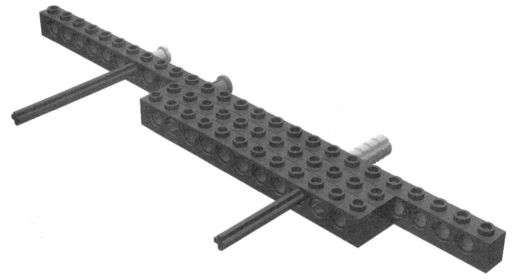

Right Drive Step 3

x1

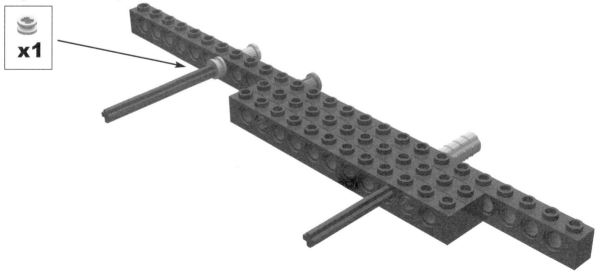

Right Drive Step 4

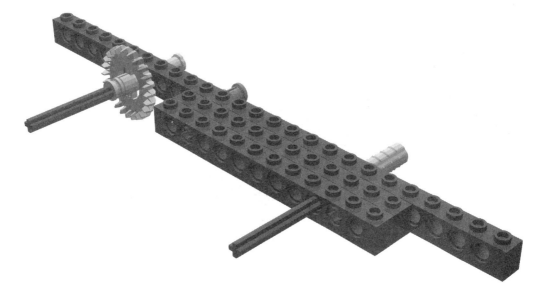

Right Drive Step 5

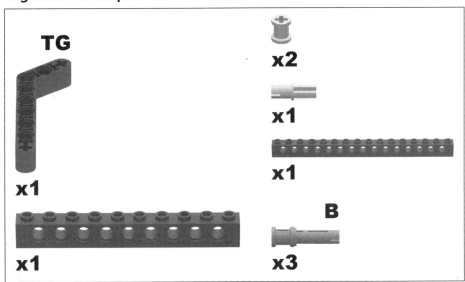

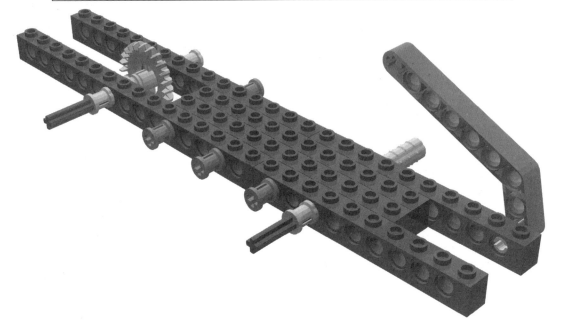

Right Drive Step 6

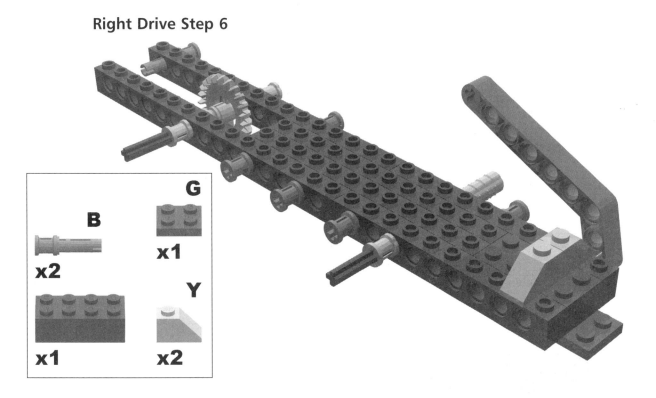

B x2
G x1
x1
Y x2

Right Drive Step 7

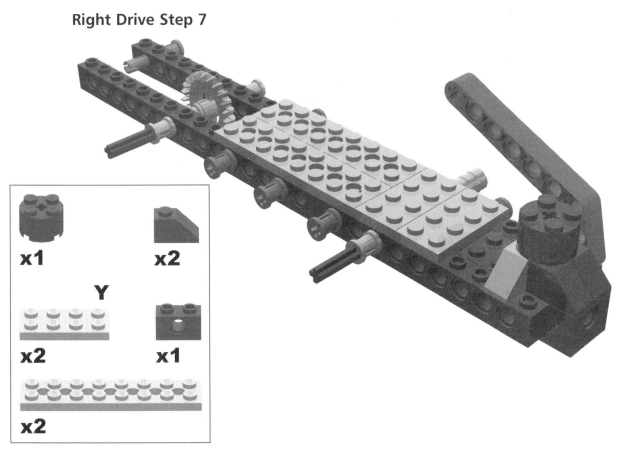

x1
x2
Y x2
x1
x2

Right Drive Step 8

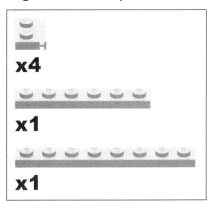

x4

x1

x1

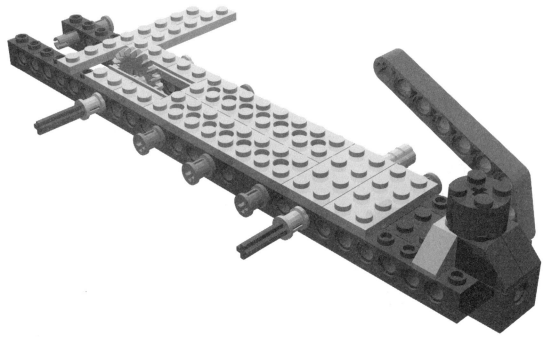

Right Drive Step 9

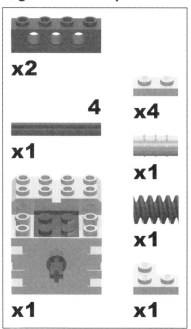

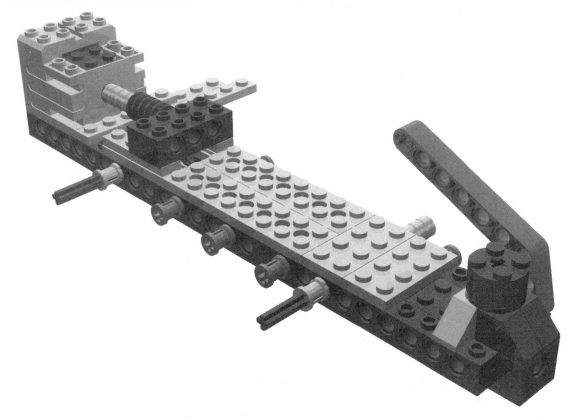

Right Drive Step 10

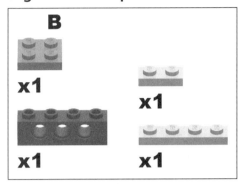

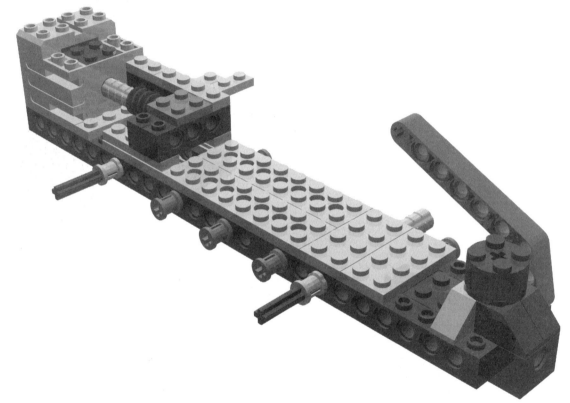

Right Drive Step 11

x1

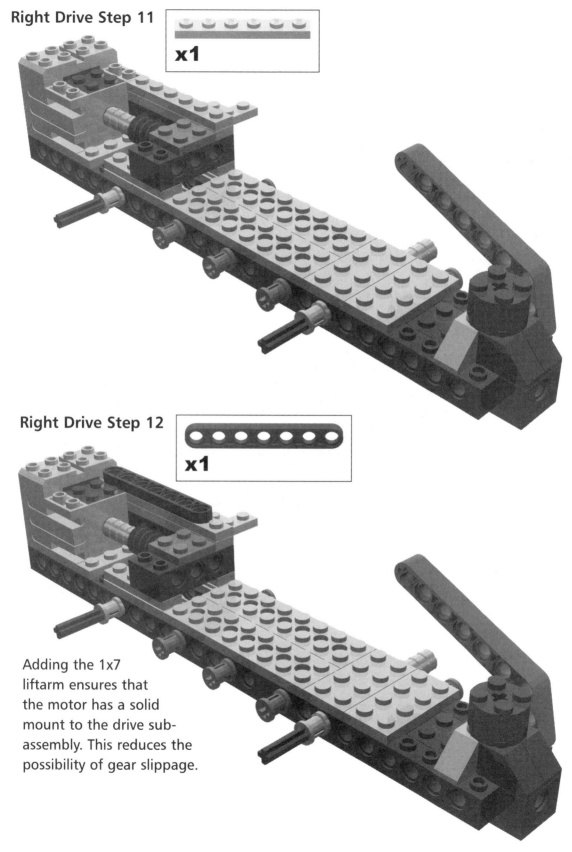

Right Drive Step 12

x1

Adding the 1x7 liftarm ensures that the motor has a solid mount to the drive sub-assembly. This reduces the possibility of gear slippage.

Right Drive Step 13

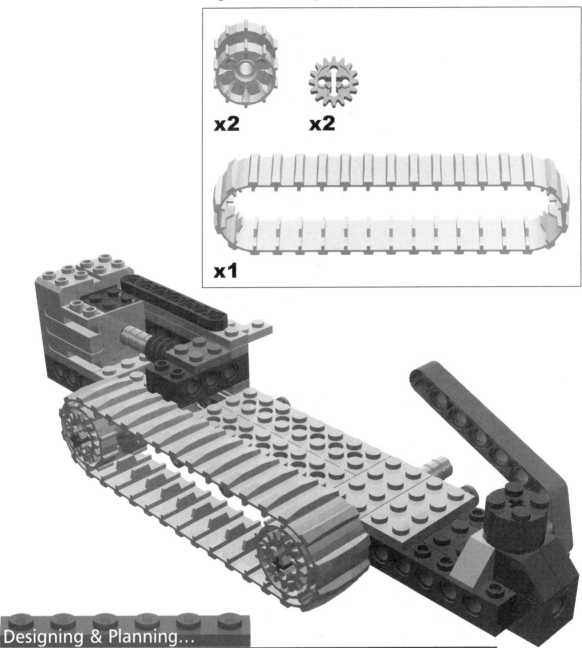

Designing & Planning...

Planning Mobile Robots

When planning a robot that requires mobility, it is important to determine the type of movement needed. Will the robot need to turn quickly? Will it need to have traction? Will it need to move fast? The mobility needs of the DominoBot required that it turn on a center rotation point like a tank. For this type of turn, tracks are used. Tracks are also useful because they allow a single hub to drive the whole unit. This is an important feature for DominoBot since the hub at the other end is used to drive the distance sensor.

The Distance Sensor

The distance sensor allows the DominoBot to track how far forward it has moved. The program will need this information in order to place the dominoes an equal distance from each other.

The differential enables the mechanism to register a click only when both the left and right drive units are moving forward. If either drive is in reverse (when steering or turning), the differential will keep the cam gear idle.

Distance Sensor Step 0

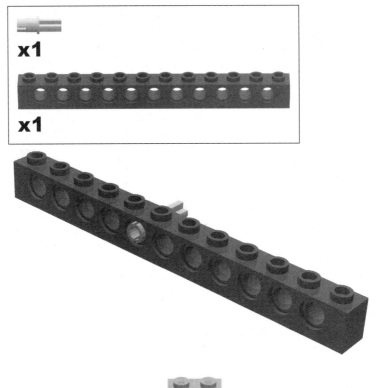

Distance Sensor Step 1

The touch sensor plays a role in the DominoBot NQC program. It is set up as a pulse sensor and registers a pulse each time the cam touches it. These pulses are counted, and a domino is placed for every third pulse.

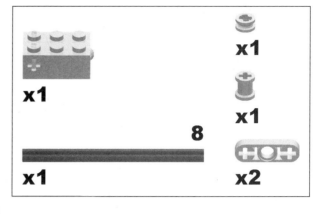

x1

x1

8

x1

x1

x2

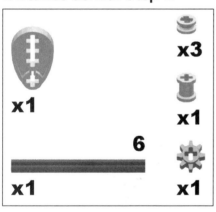

Distance Sensor Step 2

x1

x3

6

x1

x1

x1

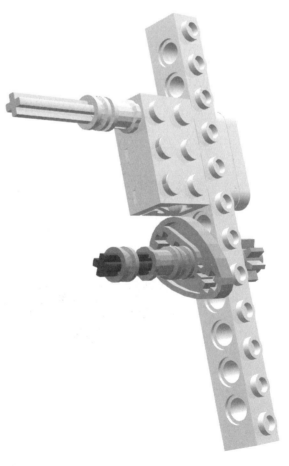

Distance Sensor Step 3

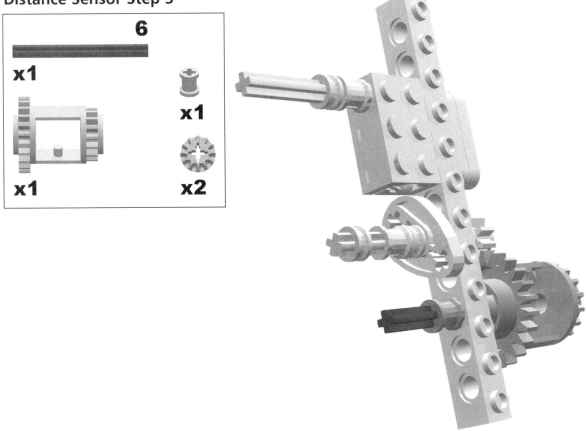

The Left Drive

The left drive sub-assembly connects to the right drive sub-assembly to form the driving base of the DominoBot. Or course, its purpose is to give it mobility.

Left Drive Step 0

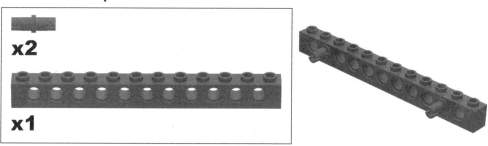

x2

x1

Left Drive Step 1

x1

Left Drive Step 2

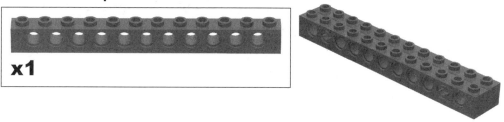

8

x1

x2

x1

x1

10

B

x1

x3

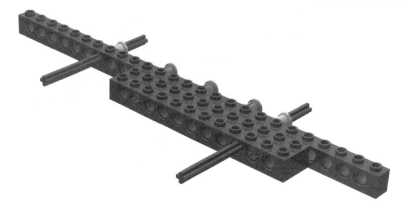

Left Drive Step 3

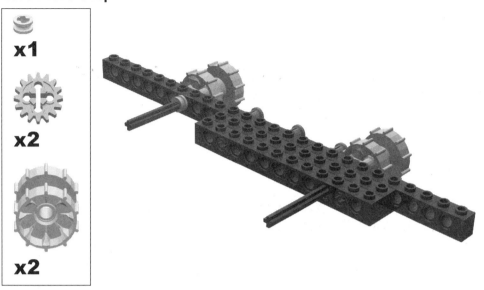

Left Drive Step 4

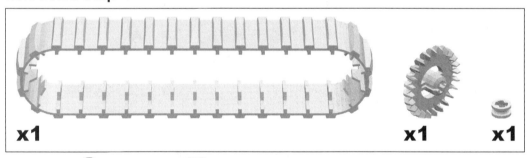

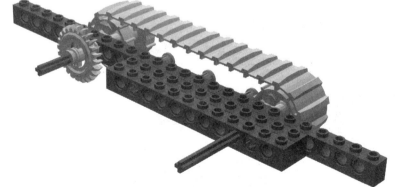

Left Drive Step 5

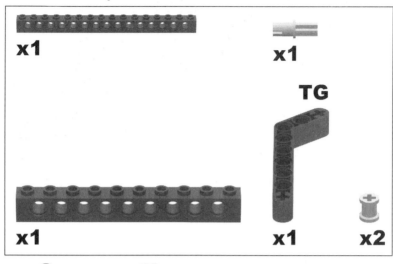

x1

x1

TG

x1

x1

x2

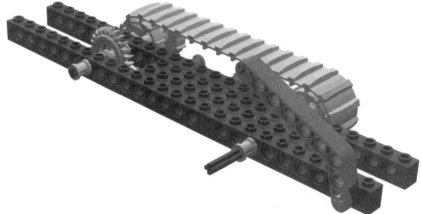

Left Drive Step 6

G

x1

x1

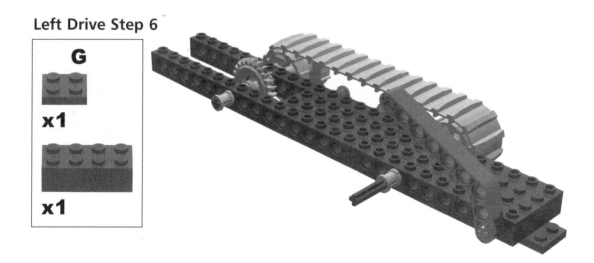

Left Drive Step 7

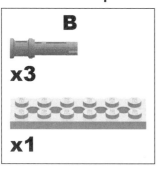

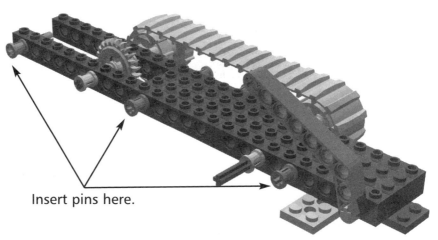

Insert pins here.

Designing & Planning...

Building to Take Things Apart

In the DominoBot, the long blue pegs with friction are used frequently in key areas. They allow for the builder to disassemble the DominoBot into its key sub-assemblies quickly. This is done for the purpose of modularity.

Left Drive Step 8

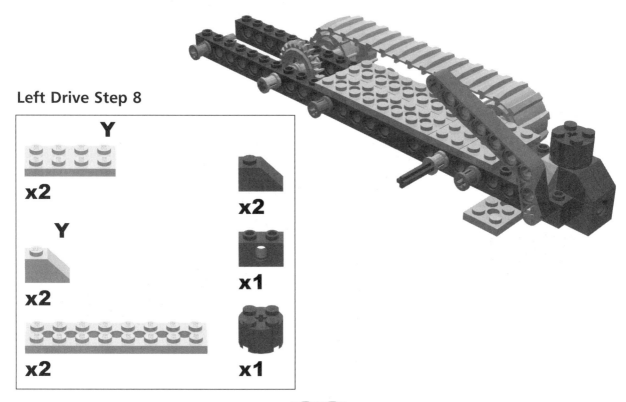

Left Drive Step 9

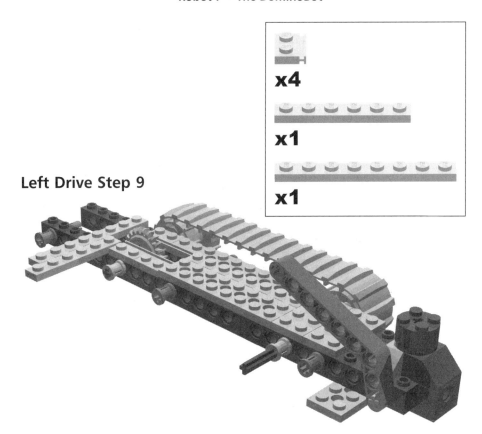

Left Drive Step 10

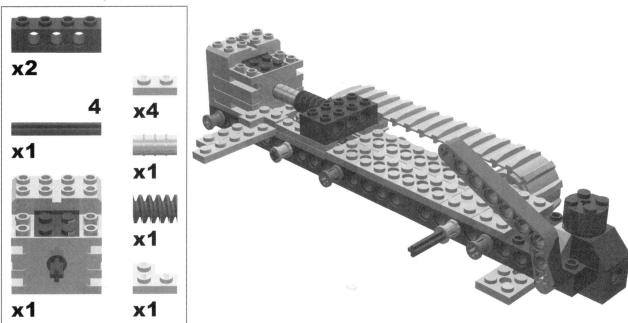

Left Drive Step 11

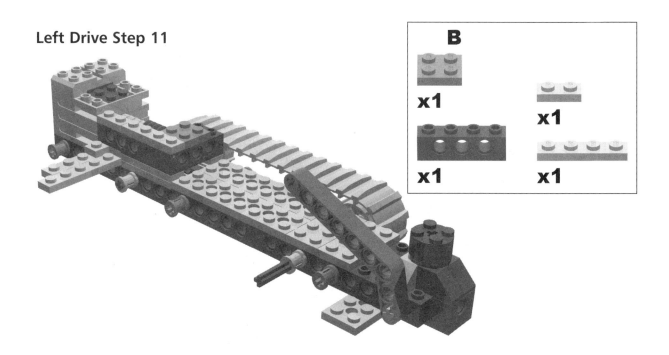

B
x1
x1
x1
x1

Left Drive Step 12

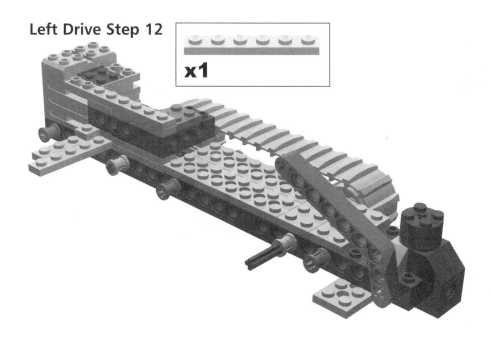

x1

Left Drive Step 13

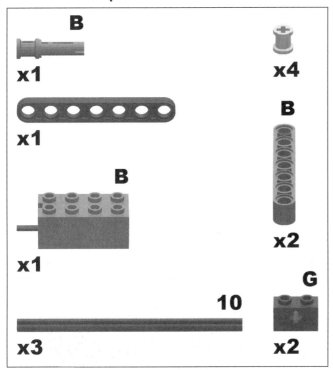

B

x1

x1

B

x1

10

x3

x4

B

x2

G

x2

Place the blue light sensor in the middle of the rear axle. It should be able to rotate around the axle. Later, when testing your DominoBot, you can rotate it to help calibrate the sensor.

Left Drive Step 14

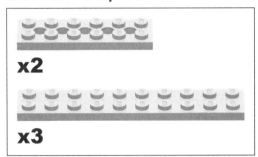

x2

x3

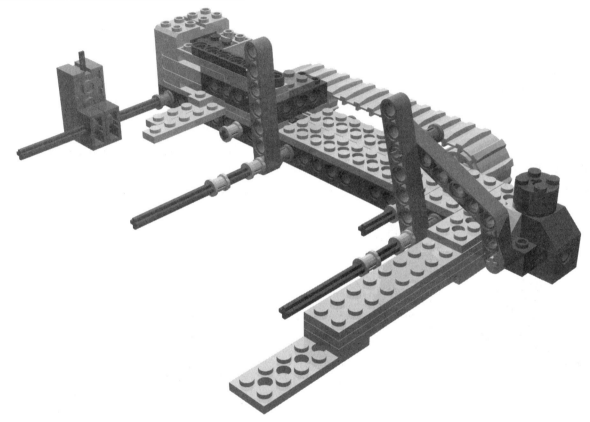

Left Drive Step 15

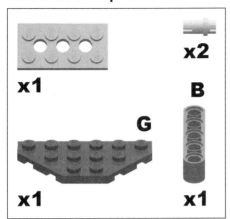

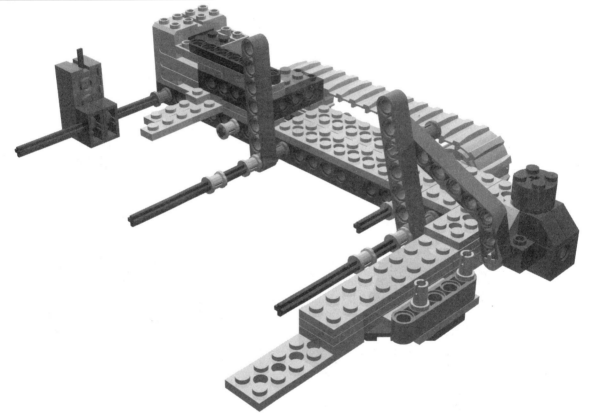

Left Drive Step 16

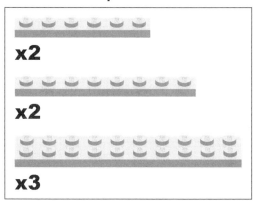

x2

x2

x3

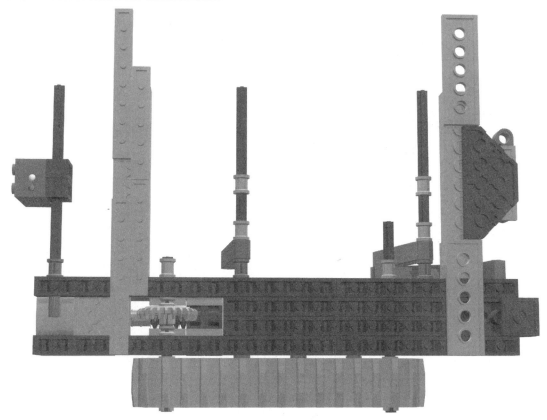

The Base

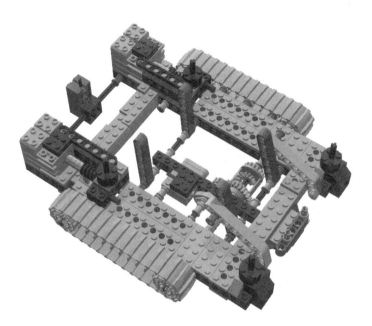

The base sub-assembly consists of the left drive sub-assembly, the right Drive sub-assembly, and the distance sensor sub-assembly. You will connect them together in the following series of steps.

Base Step 0

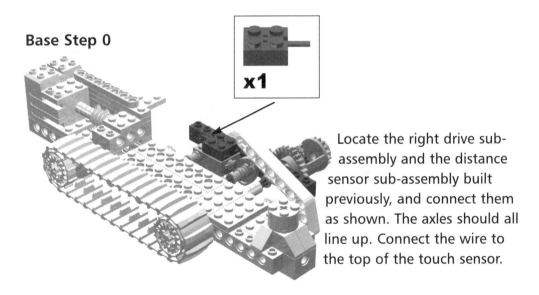

Locate the right drive sub-assembly and the distance sensor sub-assembly built previously, and connect them as shown. The axles should all line up. Connect the wire to the top of the touch sensor.

Base Step 1

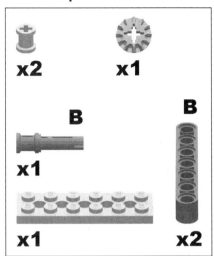

Connect the left drive sub-assembly to the structure. The order of connecting the parts is a bit complex. Start by hooking up the differential, and then inserting the 12t bevel gear.

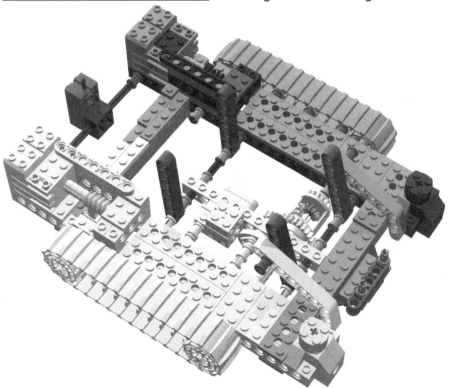

Base Step 2

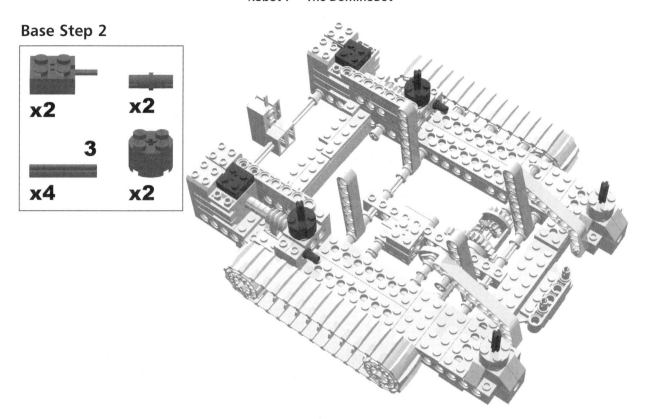

The Loader Sensor

The loader sensor sub-assembly completes the essential function of identifying to the RCX when the loader mechanism has dispensed a domino, and when the loader has reset and is ready for the next domino.

Loader Sensor Step 0

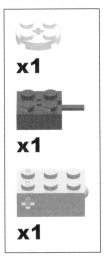

Loader Sensor Step 1

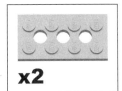

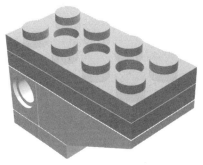

Loader Sensor Step 2

The white circular plate has an important function. In **Final Step 6**, you will insert a yellow flex tube from the loader to this plate. This will provide spring to the sensor unit to allow it to consistently maintain accurate readings while dominoes are being dispensed.

The Loader Arm

Designing & Planning...

Domino Play for LEGO

Because the DomnioBot uses standard dominoes rather than LEGO bricks, the sizes are not to LEGO specifications. In order to mimic the actions of a human hand placing a domino, an unconventional approach must be taken. The flex hoses allow for some "play" (or movement) when a domino slides into the loader arm. This play ensures that while each domino might not follow the same path down to the base of the arm, the flex tubes can straighten the placement.

Loader Arm Step 0

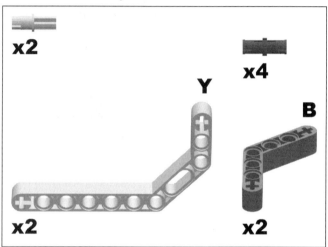

x2

x4

Y

B

x2

x2

Loader Arm Step 1

x2

x4

x2

Mount the blue perpendicular axle joiners so that one end is inserted into the yellow liftarm, while the other end floats freely.

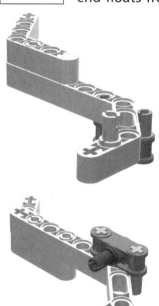

Loader Arm Step 2

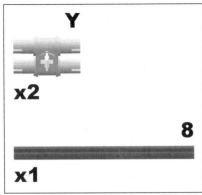

Y

x2

8

x1

Loader Arm Step 3

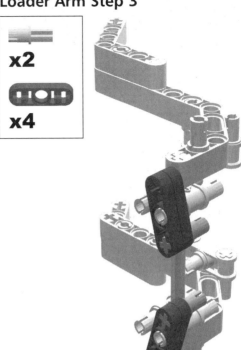

x2

x4

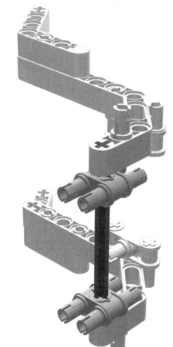

Loader Arm Step 4

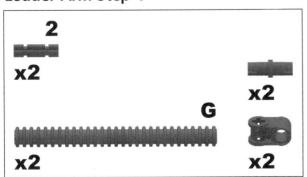

2

x2

x2

G

x2

x2

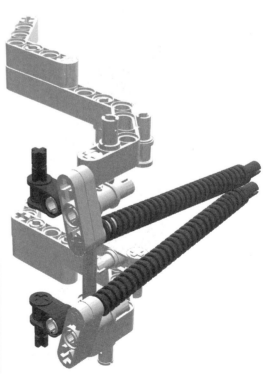

Loader Arm Step 5

x4

B

x2

Loader Arm Step 6

x1

Loader Arm Step 7

G

x2

When inserting the green flex hoses, make sure that they rest on the black axle joiners, as shown. This is important for the guidance of the domino to work correctly.

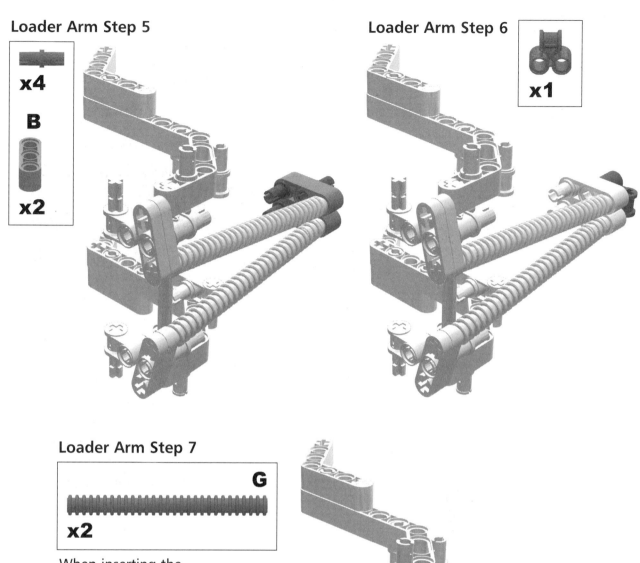

Loader Arm Step 8

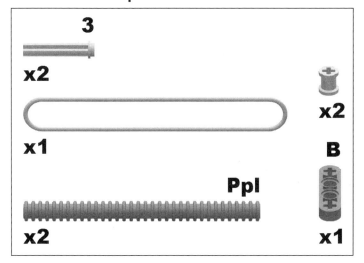

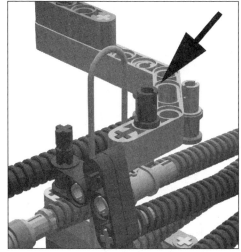

Insert the parts as shown, leaving the yellow elastic band for last. The elastic goes between the green and purple flex hoses and wraps around the black friction pegs (below) that are inserted in the yellow double-bent liftarm. Its purpose is to hold the purple hoses against the front of the loader mechanism. This is important, so study the image carefully.

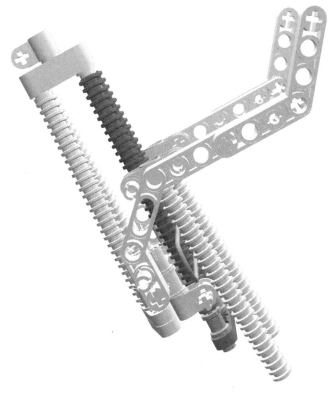

Loader Arm Step 9

Loader Arm Step 10

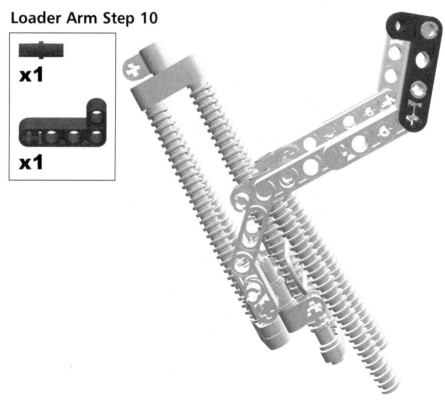

Loader Arm Step 11

Putting It All Together

Finally, put the entire DominoBot together. Here you will take all the components that you have built and put them together to complete the DominoBot!

Final Step 0

B

x4

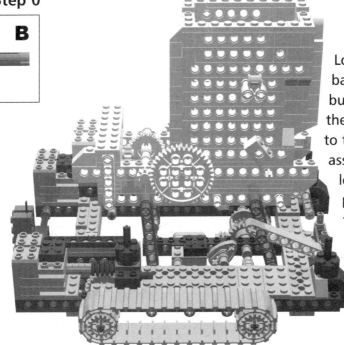

Locate the loader and base sub-assemblies built earlier. Connect the loader sub-assembly to the base sub-assembly using four, long, blue friction pegs. Take note of the location where the pegs are inserted. It is the same on both sides.

Final Step 1

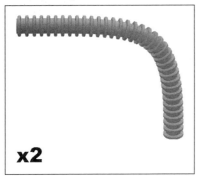

x2

Remember when you were asked to take note of the friction pins when creating the loader and the base sub-assemblies? This is where you insert the two purple flex hoses. These hoses will be the path that each domino follows when being dispensed.

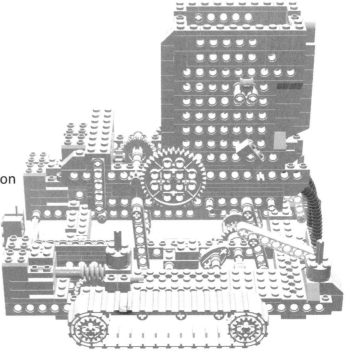

Final Step 2

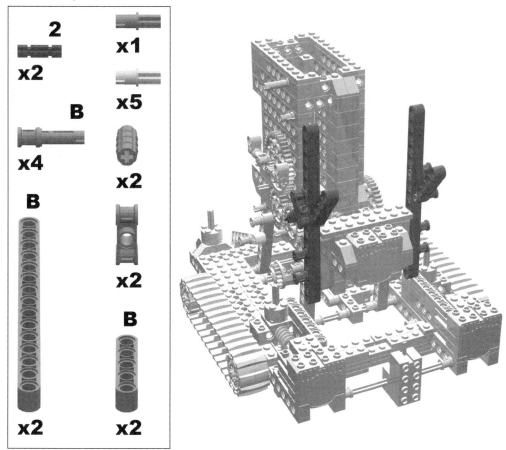

2

x2

x1

B

x4

x5

B

x2

x2

B

x2

x2

These parts will make
up the RCX holder.

Final Step 3

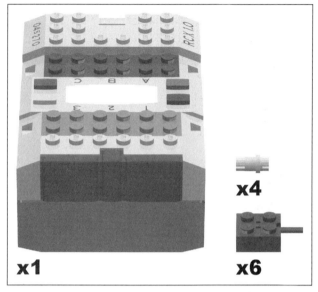

x4

x1

x6

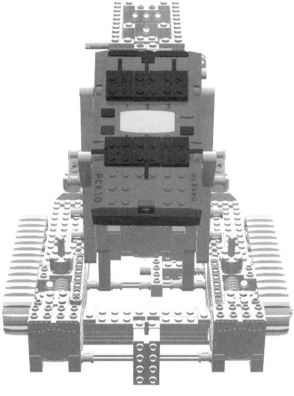

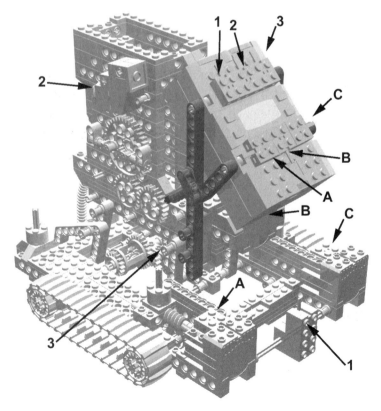

Insert the RCX. Using the diagram, match the numbers and letters. For example, RCX sensor Port 2 should go to the loader sensor Port 2 (at the top left). The light sensor has a short cable. In order for it to connect to sensor Port 1, use one of the provided medium length cables to extend its length. Connect RCX motor Port B to the loader main drive motor (below the RCX).

Final Step 4

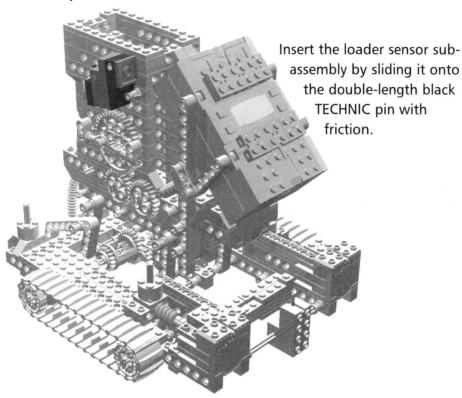

Insert the loader sensor sub-assembly by sliding it onto the double-length black TECHNIC pin with friction.

Final Step 5

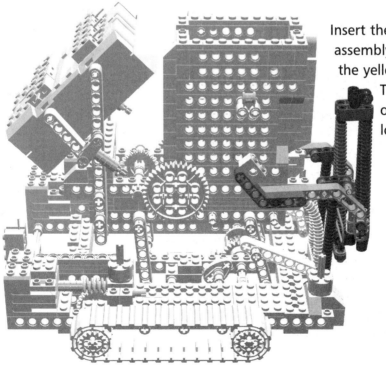

Insert the loader arm sub-assembly by connecting the yellow arms to the TECHNIC axle pins on each side of the loader.

Final Step 6

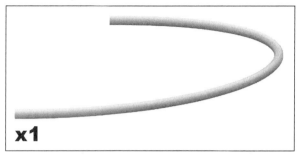

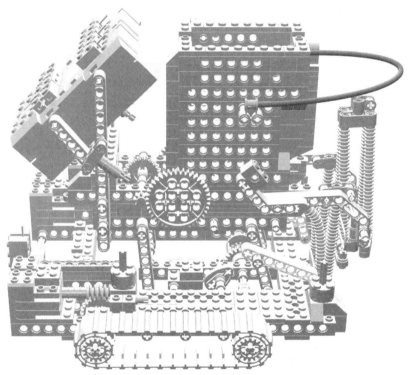

For **Final Step 6** and **Final Step 7**, take the two yellow TECHNIC flex hoses and insert them into the grey half-length pegs on the top of the loader (upper right). For the first hose (the side visible as shown here), insert the other end of the hose into the TECHNIC perpendicular double-axle joiner.

Final Step 7

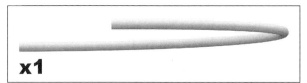

x1

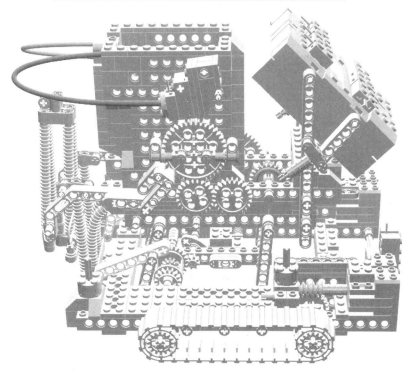

Final Step 8

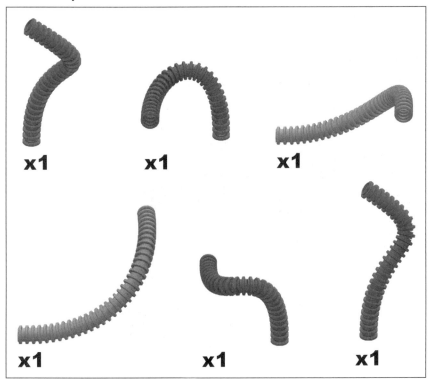

x1   x1   x1

x1   x1   x1

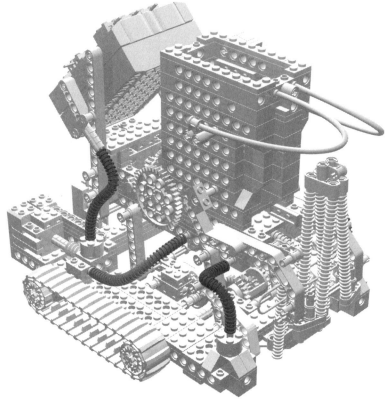

You will notice that there are some beam extensions and pegs with friction on either side of DominoBot. These pieces are the mounting points for the flexible hoses in this step. Hook up the TECHNIC flex hoses as shown here. Do the same on the opposite side, and you're done!

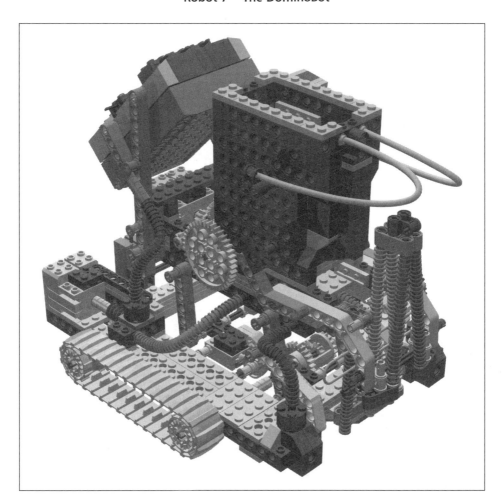

Robot 8

The Drawbridge

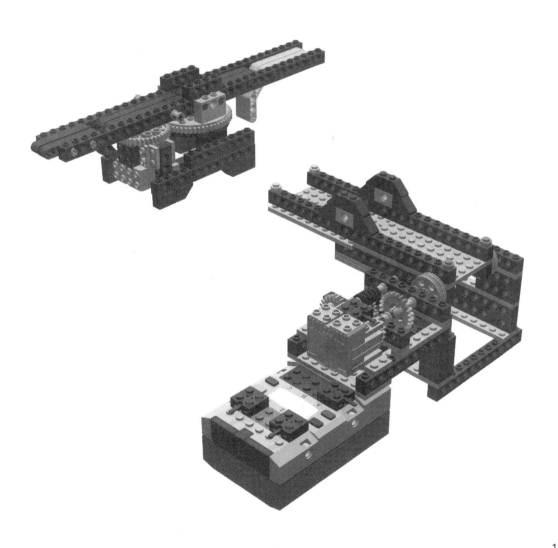

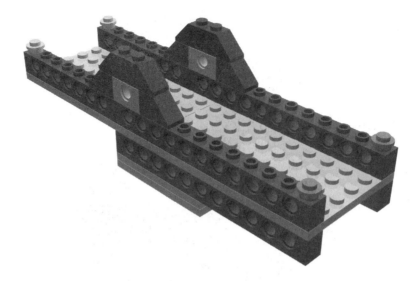

When you think of a drawbridge, you might see an image of a castle's draw-bridge spanning a moat, ready to be drawn up at a moment's notice when ene-mies appear on the horizon. Modern Drawbridges use motors to lift and lower the bridge, instead of the chains, ropes, counterweights, and pulleys that were used in the past. These days, most drawbridges are used to lift a section of road or railway so that ships can travel into a bay or down a river.

Frequently, modern drawbridges do not span an entire waterway. Instead, they connect two extension bridges to one another. The portion of the bridge that is actually "drawn up" is typically located over the deepest (and therefore most navigable) portion of the waterway. You might want to make your Drawbridge the middle section of a series of extension bridges that span a wide channel.

The gear box of the drawbridge is created by using a worm gear to create a large gear reduction so that the power from the motor is increased and the speed is decreased. This allows the motor to slowly move the heavy bridge out of the way.

The Island Footbridge is an additional bridge for pedestrians. The foot-bridge is not attached to one side of the waterway but rather anchored out in the middle like an island. The footbridge rotates around on a turntable clearing the waterway so that objects may pass down the river.

SYNGRESS

syngress.com

The Drawbridge is programmed to use the light sensor to detect when a ship or boat is approaching. The RCX beeps a tone and lifts the movable por-tion, or *bascule* or *balance*. When the light sensor detects that the ship has passed by, the RCX beeps another tone and begins to lower the bascule. When the bascule is completely lowered it will press against the touch sensor, causing the motor to stop lowering the bascule and sounding another tone so that cars will know it is safe to pass over the Drawbridge. The program for the Drawbridge can be downloaded from the Syngress Solutions Web site (www.syngress.com/solutions).

Bridge Base

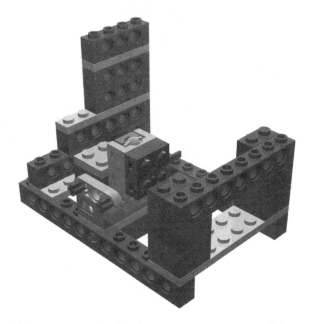

The bridge base sub-assembly supports the bridge as it moves up and down. The base of the bridge must be sturdy enough to support the weight of cars and trucks passing over it.

Base Step 0

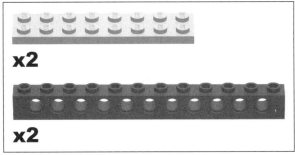

x2

x2

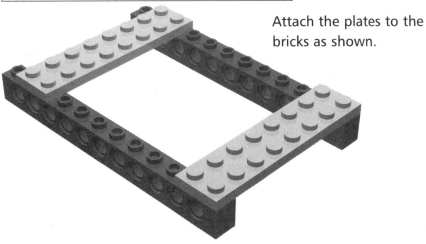

Attach the plates to the bricks as shown.

Base Step 1

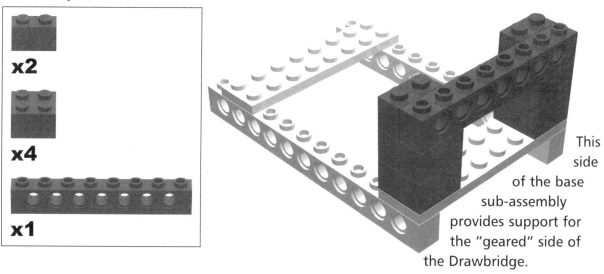

x2

x4

x1

This side of the base sub-assembly provides support for the "geared" side of the Drawbridge.

Base Step 2

This side of the base sub-assembly provides support for the opposite "ungeared" side of the Drawbridge.

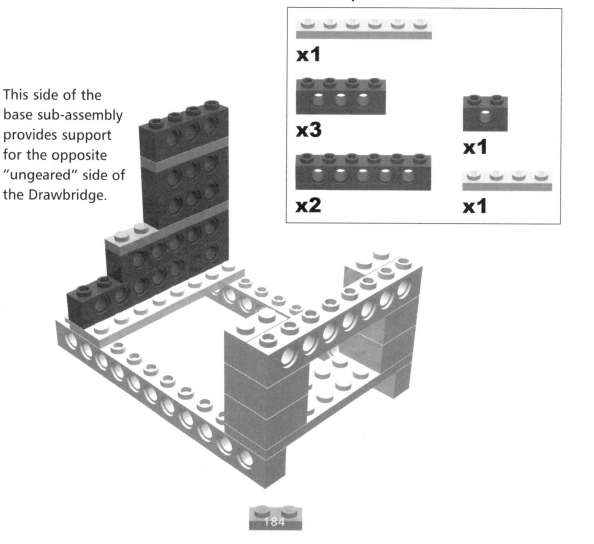

x1

x3

x1

x2

x1

Base Step 3

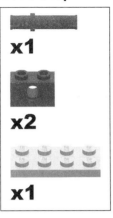

This step provides the base for the sensors. The 1x2 TECHNIC brick is attached to the underside of the 2x8 plate. Use the long pin with the stop bushing to connect the 1x2 brick and the frame.

Base Step 4

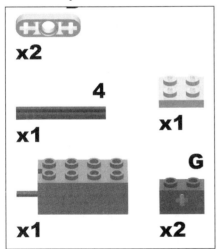

The light sensor will detect a ship, car, or train as it begins to pass under the bridge, causing the bridge to lift the bascule.

Attach the green 1x2 bricks with axle holes to the top of the light sensor. The light sensor wires should point towards the center of the frame. Layer the 2x4 brick on top of these bricks. Next, slide the #4 axle through the bricks. Attach the blue liftarms to the axle as shown.

Adding Warning Lights

Program the light sensor to detect a flashlight that acts as a warning to lift the bridge.

Base Step 5

x1

x1

4

x1

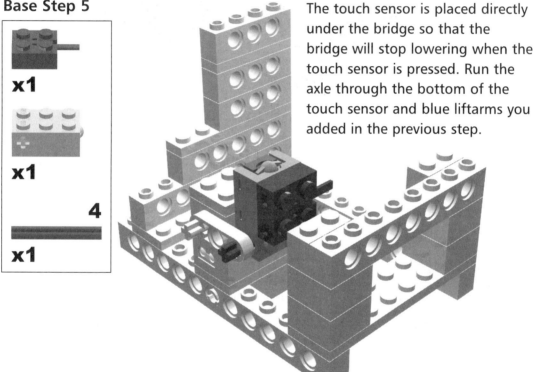

The touch sensor is placed directly under the bridge so that the bridge will stop lowering when the touch sensor is pressed. Run the axle through the bottom of the touch sensor and blue liftarms you added in the previous step.

Touch Sensor Placement

Try placing the touch sensor on the end of the bridge to stop the motor when the bridge contacts the far support.

Motor Assembly

The motor assembly includes a gearbox with two worm gears that provide low speed and high torque, which is exactly what is needed to lift the heavy bascule.

Bricks & Chips...

Worm Gears: How Do They Work?

Worm gears are used when large gear reductions are needed. Gearing down provides high power at low speed. The worm gear drives the gear, but the gear cannot drive the worm. This locking feature can act like a brake.

Motor Assembly Step 0

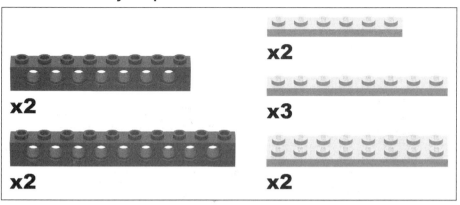

This step begins by building a stable frame for the gearbox. Arrange the bricks and plates as shown, building from the bottom up.

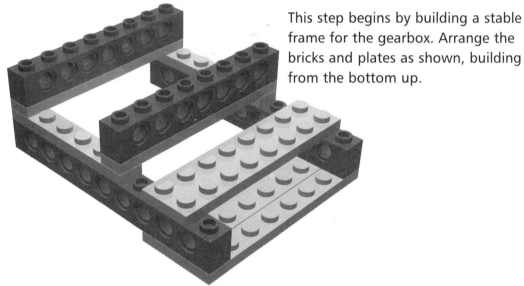

Motor Assembly Step 1

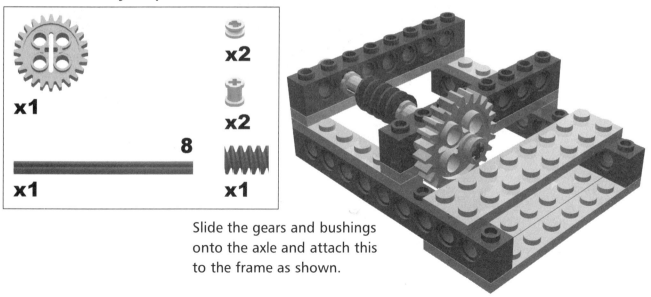

Slide the gears and bushings onto the axle and attach this to the frame as shown.

Motor Assembly Step 2

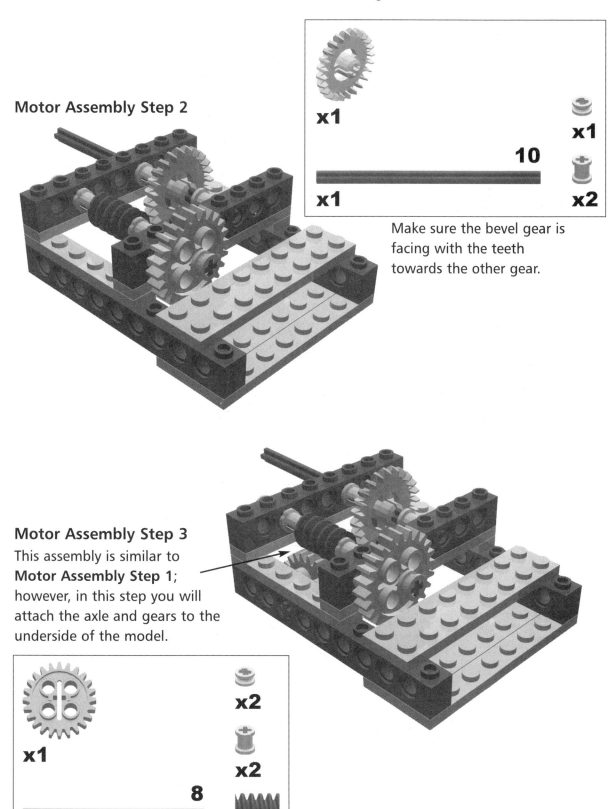

x1

10

x1

x1

x2

Make sure the bevel gear is facing with the teeth towards the other gear.

Motor Assembly Step 3

This assembly is similar to **Motor Assembly Step 1**; however, in this step you will attach the axle and gears to the underside of the model.

x1

x2

x2

8

x1

x1

Motor Assembly Step 4

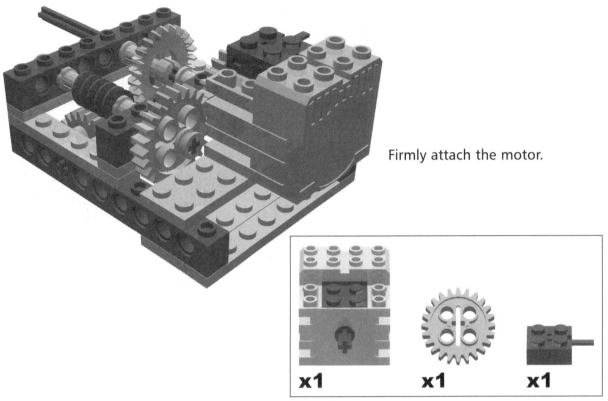

Firmly attach the motor.

x1   x1   x1

Motor Assembly Step 5

x2

x1

x2

Adding the axle extender lengthens the short axle leading from the motor.

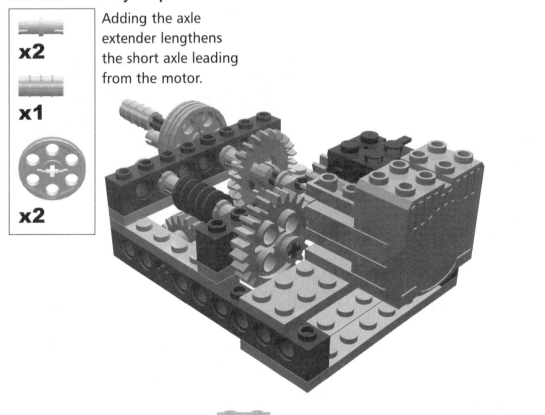

Motor Assembly Step 6

The position of the pins enables the movement from the gearbox to be translated to the bridge.

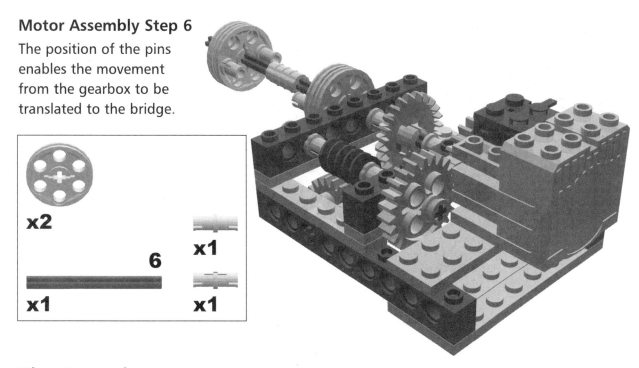

The Bascule

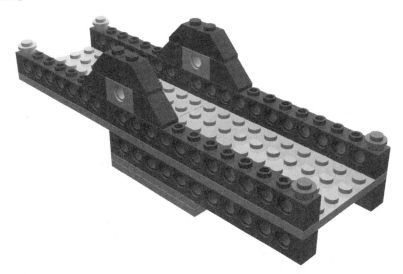

The Bascule sub-assembly is the moveable part of the bridge that will lift or draw. It is often also called a *balance*.

Inventing...

Lengthening the Bascule

This same pattern can be used to extend the bascule even further (be careful, as the gears can only lift so much weight!).

Bascule Step 0

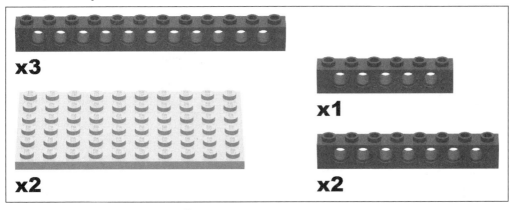

x3

x1

x2

x2

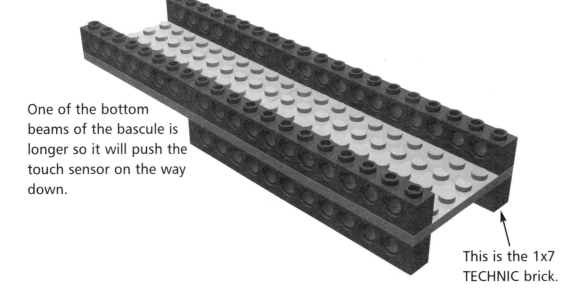

One of the bottom beams of the bascule is longer so it will push the touch sensor on the way down.

This is the 1x7 TECHNIC brick.

Bricks & Chips...
Adding Personal Touches

You can add flags or suspension wires to the bascule as desired.

Bascule Step 1

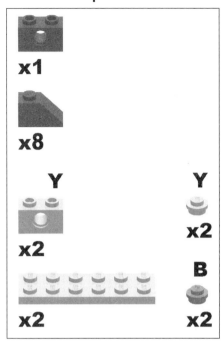

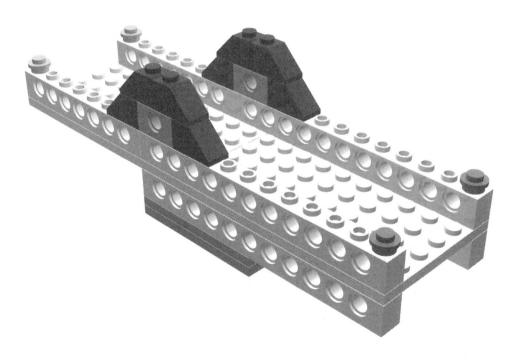

The Island Footbridge

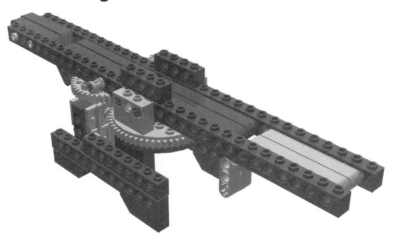

The island footbridge sub-assembly is anchored in the middle of the waterway like an island. The island footbridge rotates on a turntable to move out of the way to let even larger types of boats pass.

Island Footbridge Step 0

x6

x2

Place three connectors in the holes of each of the beams as shown.

Island Footbridge Step 1

x1   x1

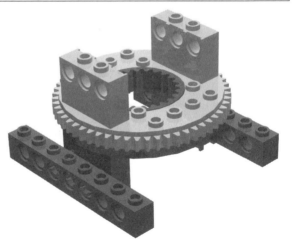

Connect the turntable to the beams.

Island Footbridge Step 2

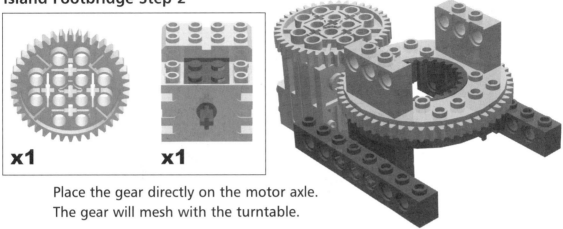

Place the gear directly on the motor axle.
The gear will mesh with the turntable.

Island Footbridge Step 3

Island Footbridge Step 4

x2

x4

x2

x4

These parts will create the supports for the island footbridge on the turntable.

Island Footbridge Step 5

x4

x4

Island Footbridge Step 6

Although not visible in this step, add an electrical wire to connect the motor to the RCX. The motor will turn on when the light sensor sees a boat passing. This will cause the turntable to rotate and clear the waterway.

x1

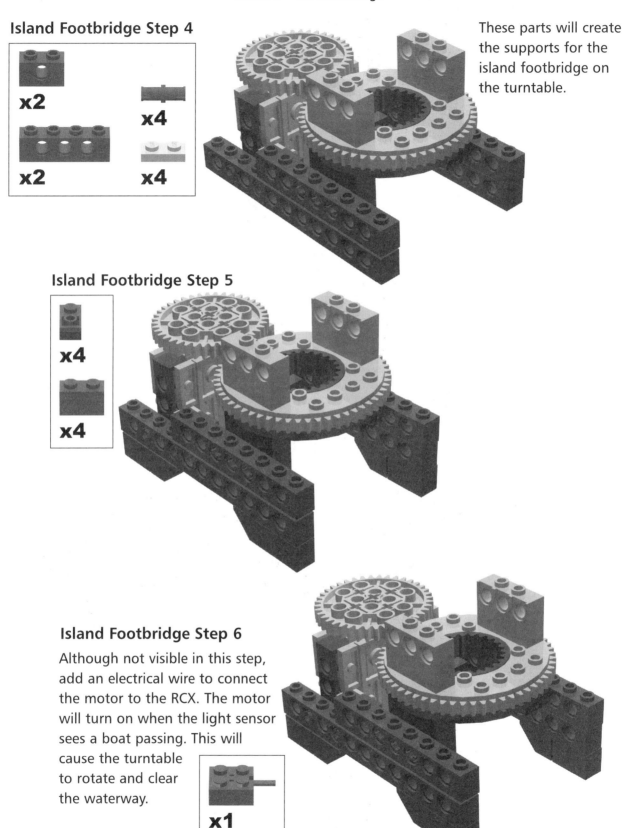

Bricks & Chips...

Long Electrical Wires

Add a long motor wire to connect to the RCX. This will allow the Island Footbridge to be placed far enough away from the drawbridge so that they don't collide.

Island Footbridge Step 7

x4

x4

Attach the beams that form the sides of the footbridge.

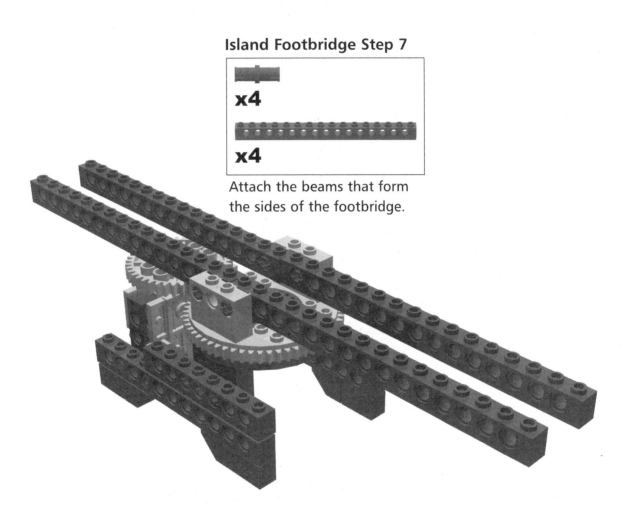

Island Footbridge Step 8

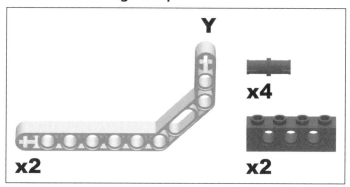

Add the beams to form the floor of the footbridge.

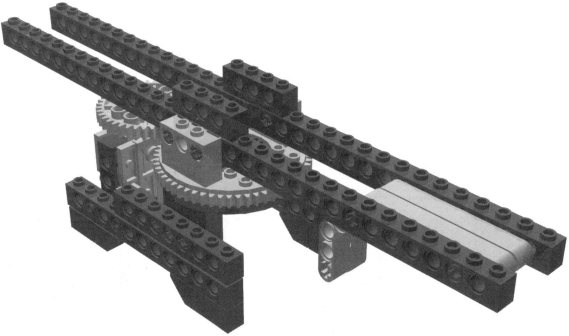

Island Footbridge Step 9

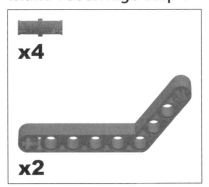

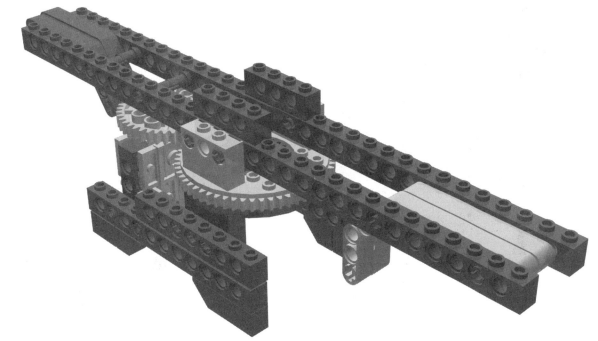

Island Footbridge Step 10

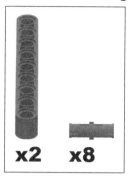

x2   **x8**

Continue adding beams to build up the footbridge.

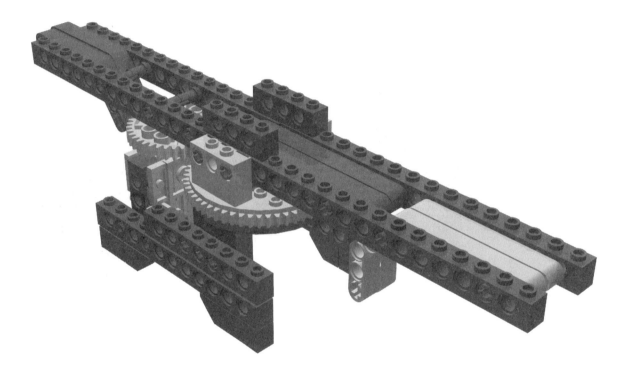

Island Footbridge Step 11

x4

6

x1

Run a #6 axle
through the bridge
and secure it with the
bushings as shown.

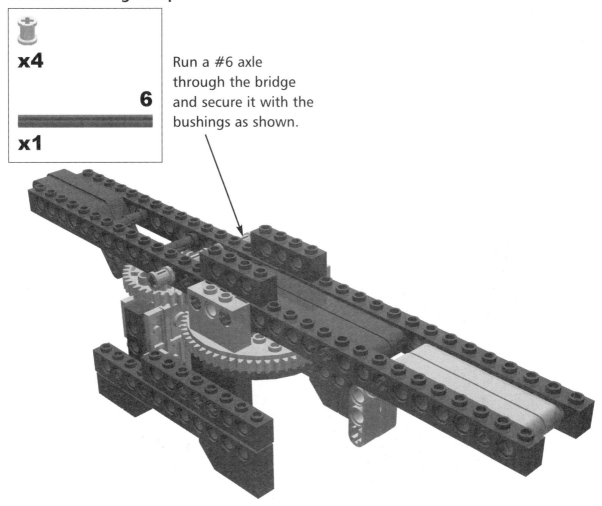

Island Footbridge Step 12

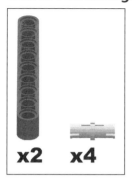

x2   x4

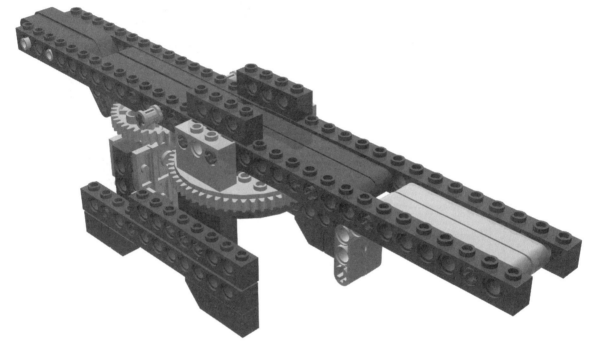

Inventing...

Building a Bigger Bridge

Another way to lengthen the bridge is to build a non-moveable extension bridge segment on either side of the movable bascule.

Putting It All Together

The final steps include addition of the RCX, and snapping together of all of the previously built sub-assemblies. The RCX controls the movement of the Drawbridge and Island Footbridge, and sounds warning tones to let others know when the Drawbridge is lifting, lowering, and when it is safe again for traffic to cross.

Developing & Deploying...
Lengthening Electric Wires

Attach a short wire to the end of the light sensor wire. This will lengthen the electrical wire so that it will reach the RCX. All of the electrical wires can be attached end to end to continue the electrical connection.

Final Step 0

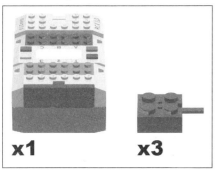

x1 **x3**

Attach the wires leading from the light sensor to Input Port 1. Attach the wires leading from the touch sensor to Input Port 3. Attach the wires leading from the motors to Output Port A and B.

Final Step 1

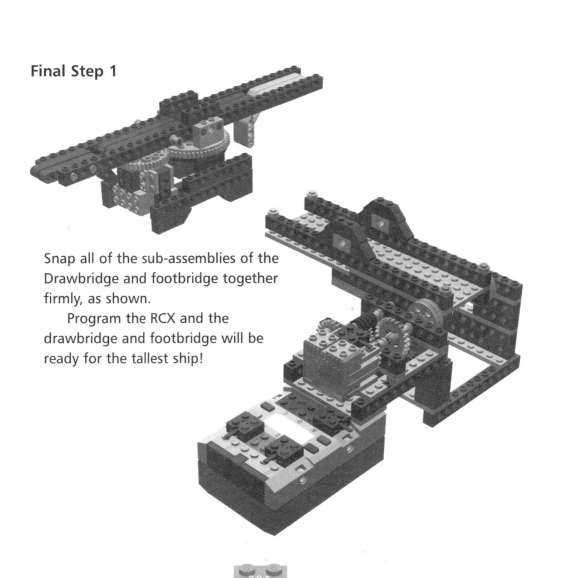

Snap all of the sub-assemblies of the Drawbridge and footbridge together firmly, as shown.

Program the RCX and the drawbridge and footbridge will be ready for the tallest ship!

Robot 9

Candy Wrapper Compactor

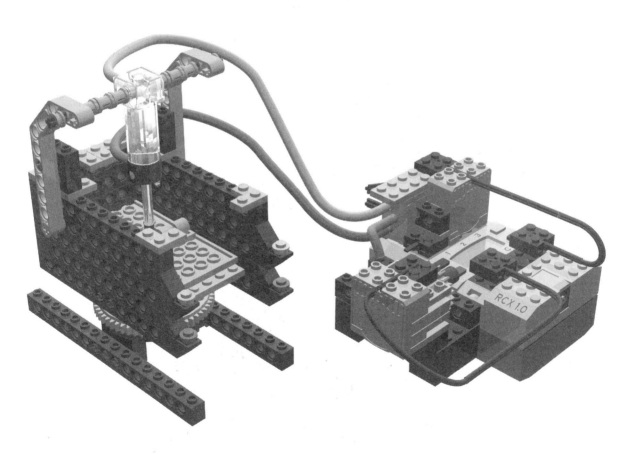

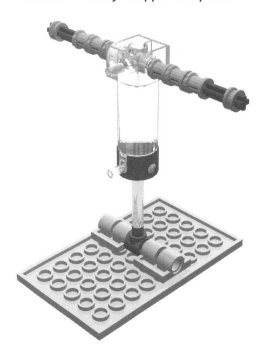

Unlike the trash compactor you may have in your home, the Candy Wrapper Compactor is made for small pieces of trash–like candy wrappers.

Compactors compress waste so that two to three times as much trash fits in the same size space. This saves storage space, removal time, and disposal labor. Special compactors called balers are used to bale or bundle recyclable material such as cardboard, paper, plastics, metal, and so on for resale to a recycling company. Compactors and balers are especially helpful in compacting waste on airplanes, ships, and restaurants. Compactors are also used to pack hazardous materials into drums for safer removal.

The Candy Wrapper Compactor is made so that you can eat a bag of Hershey's kisses at your desk, or in your room, and compact all of the wrappers into one small and neat bale that can easily be recycled or tossed into the wastebasket. This will take care of the problem of little wrappers collecting everywhere during snack time.

This is how the Candy Wrapper Compactor works:

- The compactor will start when the touch sensor (start button) is pressed.

- The compressor motor will start and move the pump up and down pushing air through the tubing.

- The RCX Compressor Control motor will turn on and move the valve allowing the air to move through the tubing engaging the piston.

- The piston will fire, which will push the ram plate down.

- When the ram plate descends, it compresses the candy wrappers.

- The RCX Compressor Control motor then reverses direction causing the axle to hit the valve changing the airflow from the compressor so that the piston pulls the ram plate back up and ends the compacting process.

The pneumatic compressor system can be used to supply pressurized air to supply power for other robots or projects. As the pump begins to move, the pressure builds and provides power. Air tanks can be added to the system to store the air so there is no delay in the supply of air power. For projects that require more than one or two pumps to power additional movements, air power can be supplied by adding a duplicate compressor on the other side of the RCX.

The Bin

The bin holds items to be compacted. The sides contain the waste so that the ram can compress the waste.

Bin Step 0

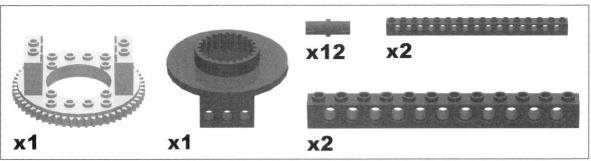

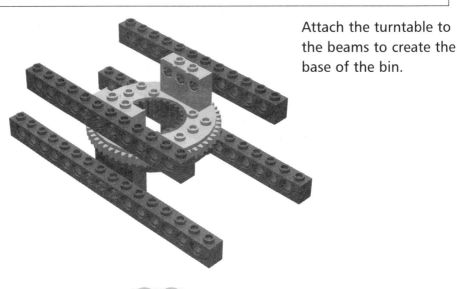

Attach the turntable to the beams to create the base of the bin.

Bin Step 1

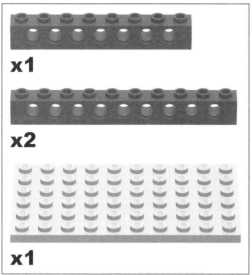

x1

x2

x1

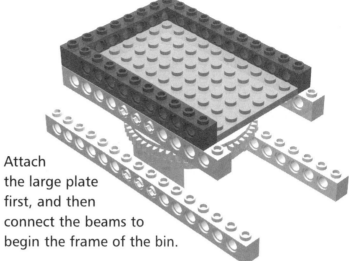

Attach the large plate first, and then connect the beams to begin the frame of the bin.

Bin Step 2

x2

x1

x2

x2

B

x2

Add beams to build up the frame.

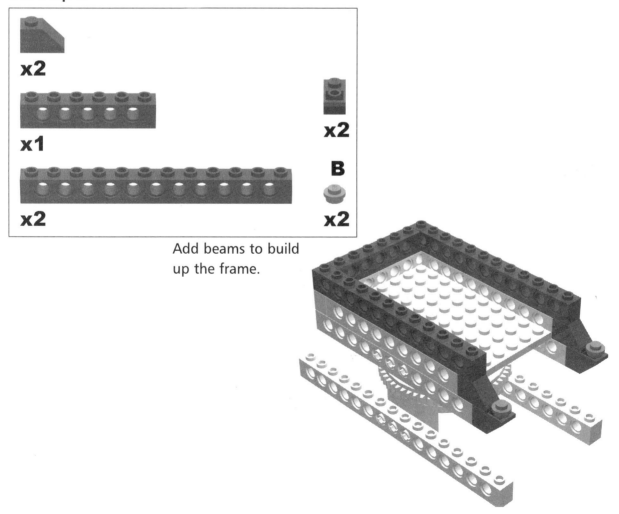

Bin Step 3

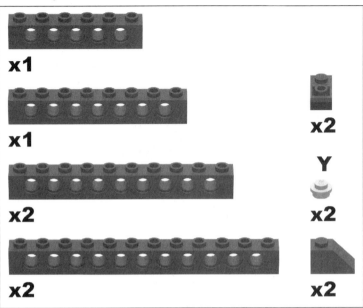

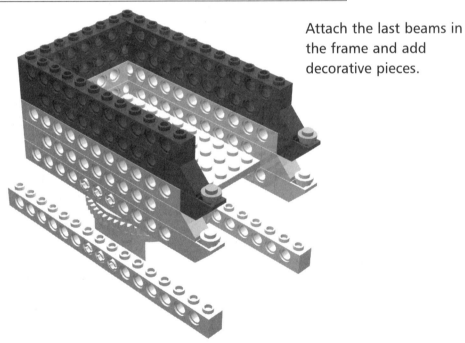

Attach the last beams in the frame and add decorative pieces.

Bin Step 4

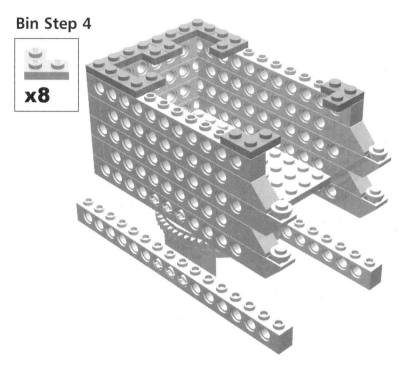

x8

Add the corner pieces to guide the ram as it moves up and down. These corner pieces work with the 2x2 bricks that you have the option of adding in **Pneumatic Ram Step 0**.

Bin Step 5

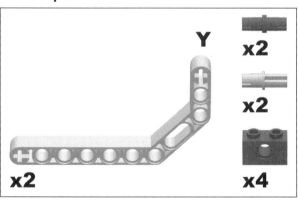

Y

x2

x2

x2

x4

Connect the yellow liftarms to create the structure that holds the pneumatic ram sub-assembly. Add the 1x2 bricks for additional support.

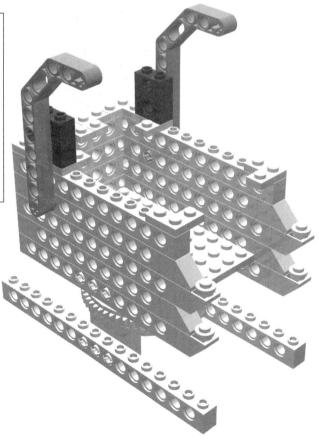

The Pneumatic Ram

The pneumatic ram sub-assembly consists of a plate and a pneumatic piston that moves up and down to compress waste. In our case the "waste" will be our candy wrappers.

Pneumatic Ram Step 0

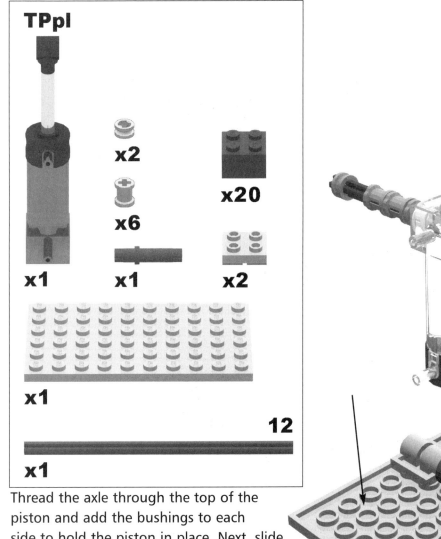

Thread the axle through the top of the piston and add the bushings to each side to hold the piston in place. Next, slide the long pin through the bottom of the piston. Slide the 2x2 plates with connectors onto each end of the long pin. And, connect the piston to the base plate as shown.

An additional way to ensure that the ram moves straight up and down is to add a column of five 2x2 bricks to each corner of the ram plate. These columns should be flush with the corners of the plates, this is completed by attaching the columns directly into the circles on the plates, rather than using the spaces between the circles as you typically would.

Bricks & Chips...

Pneumatic Ram Stability

The addition of the four columns of 2x2 bricks to each corner of the plate is a suggestion for stability and is not shown in the step image. However, we urge you to try both methods to determine which one works best for you and the types of wrappers you like to crush.

Bricks & Chips...

Having Trouble?

You might find it easier to attach the base plate in this step to the bin prior to attaching the piston.

The RCX Air Compressor Control

The RCX Air Compressor Control sub-assembly controls the flow of air. The RCX is programmed to power the motor when the touch sensor is pressed.

Inventing...

Programming Ideas

Program the touch sensor to turn the compressor motor on to start the Candy Wrapper Compactor. Make sure to write the program to have the start action of the touch sensor correspond with the Input Port 1.

RCX Air Compressor Step 0

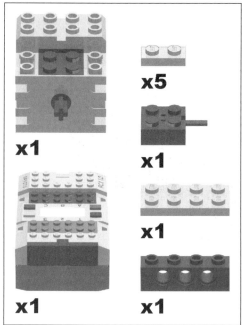

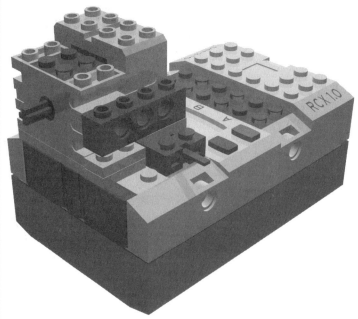

Connect the 2x4 plate to the RCX first to create a level platform for the motor. Next, connect the motor and the electrical wire to RCX Input Port 1. This wire will be attached to the touch sensor in a later step. Finish this step by attaching the 1x2 plate which will provide a level surface to attach the 1x4 TECHNIC brick.

RCX Air Compressor Step 1

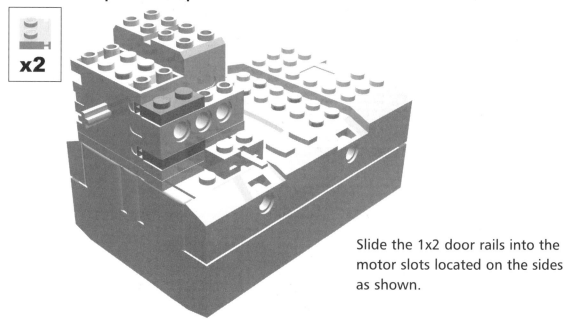

Slide the 1x2 door rails into the motor slots located on the sides as shown.

RCX Air Compressor Step 2

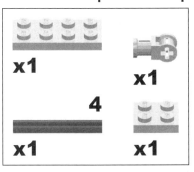

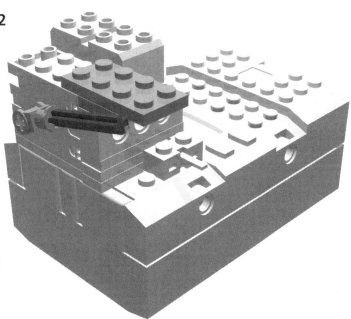

Slide the axle into the axle connector and then attach both pieces to the motor. Add plates to secure the motor.

RCX Air Compressor Step 3

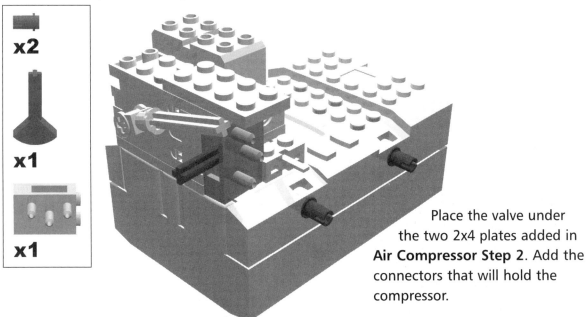

Place the valve under the two 2x4 plates added in **Air Compressor Step 2**. Add the connectors that will hold the compressor.

Bricks & Chips...

The Pneumatics

The axle in **RCX Air Compressor Step 2** will be the mechanism that strikes the pneumatic valve. The axle on the motor will shift the valve and airflow which is needed to lift the pneumatic ram sub- assembly up or push the sub-assembly down.

The Compressor

The compressor sub-assembly consists of a motor with a gear. This will move the pump in and out to create air pressure; the air will then be pushed through the tubing to move the piston.

Compressor Step 0

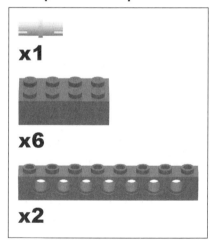

Connect the beams to the bricks to build the frame.

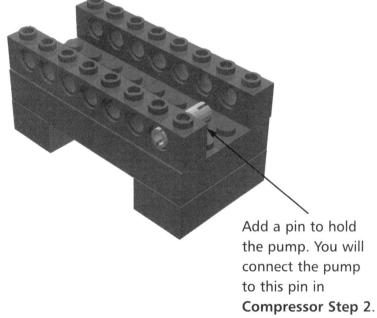

Add a pin to hold the pump. You will connect the pump to this pin in **Compressor Step 2**.

Compressor Step 1

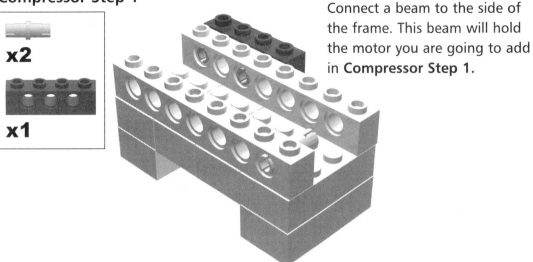

Connect a beam to the side of the frame. This beam will hold the motor you are going to add in **Compressor Step 1.**

Compressor Step 2

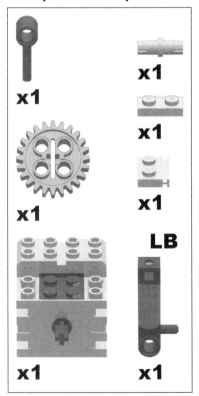

x1

x1

x1

x1

x1

LB

x1

x1

Slide the 1x2 flat plate with a door rail into the side of the motor and attach the motor to the frame. Next, add the 24t gear to the axle of the motor. Insert a pin into the holes in the gear. Then, slide a connector onto the the pump. Finally, connect the pump to the pin on the gear added in this step, and the pin in the frame added in **Compressor Step 0.** You may have to slightly bend the pump to maneuver the pit into the compressor assembly. Be careful not to break the pump.

Putting it All Together

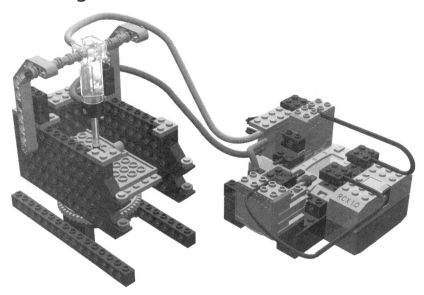

Connect the pneumatic tubing, electrical wires and all of the finished assemblies to complete the Candy Wrapper Compactor.

Final Step 0

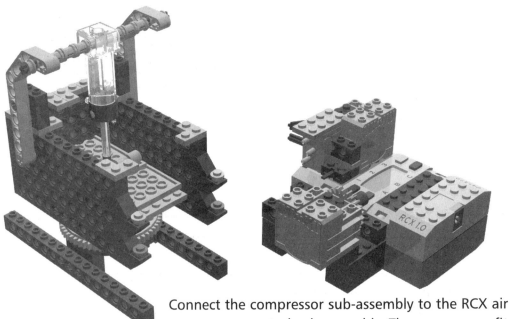

Connect the compressor sub-assembly to the RCX air compressor control sub-assembly. The compressor fits into the connectors on the side of the RCX.

You will also connect the axle on the pneumatic ram sub-assembly into the front most holes of the yellow liftarms in the bin sub-assembly.

Final Step 1

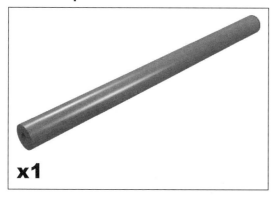

x1

Connect one end of a pneumatic tube to the middle of the compressor sub-assembly and the other end to middle input of the valve on the RCX air compression control sub-assembly. In this step we have used a customized short tube.

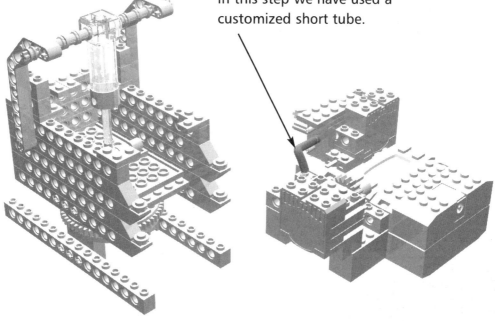

Bricks & Chips...

Tubing Length

Many builders find usually find that long tubes in some assemblies are cumbersome. It is quite simple to snip a piece of tubing to the length that you need. However, if you do not want to cut your tubing, a longer tube will work just as well.

Final Step 2

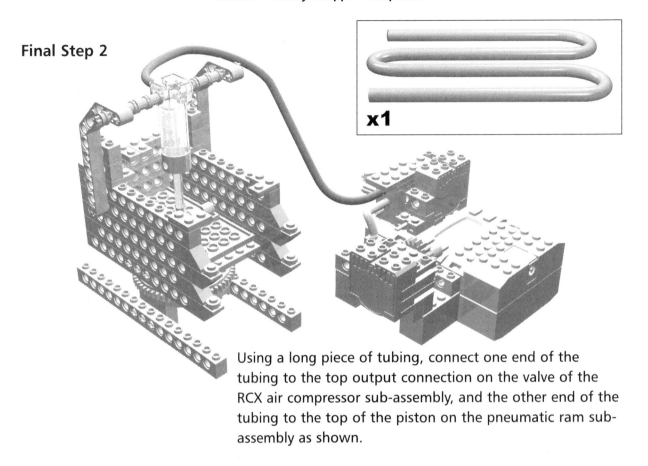

x1

Using a long piece of tubing, connect one end of the tubing to the top output connection on the valve of the RCX air compressor sub-assembly, and the other end of the tubing to the top of the piston on the pneumatic ram sub-assembly as shown.

Final Step 3

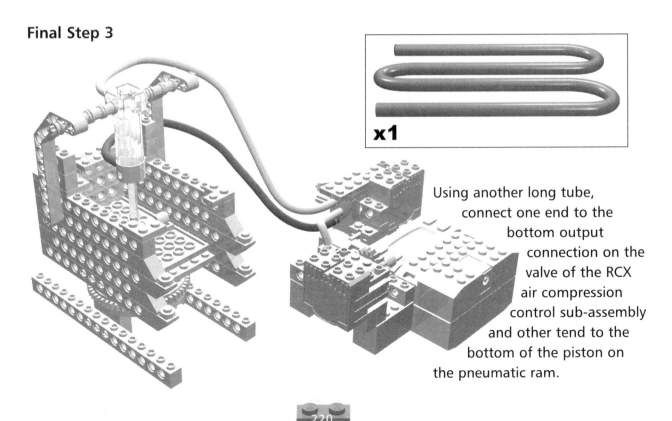

x1

Using another long tube, connect one end to the bottom output connection on the valve of the RCX air compression control sub-assembly and other tend to the bottom of the piston on the pneumatic ram.

Final Step 4

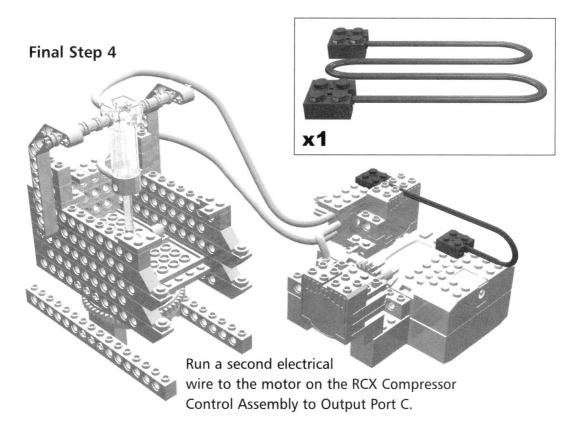

Run a second electrical
wire to the motor on the RCX Compressor
Control Assembly to Output Port C.

Final Step 5

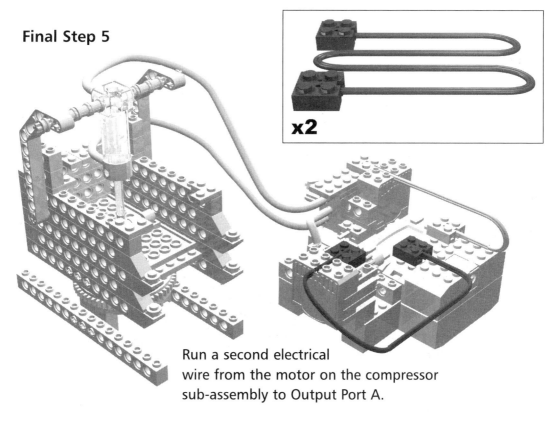

Run a second electrical
wire from the motor on the compressor
sub-assembly to Output Port A.

The Start Button

Remember that wire that you added to your robot in **RCX Air Compressor Step 0**? At this point your robot should have that free hanging wire.

The Touch Sensor

Attach the end of the electrical wire connected to Input Port 1 to a touch sensor.

The touch sensor is the start button for the Candy Wrapper Compactor. The touch sensor can be placed anywhere on the bin sub-assembly or even held in your hand as a wired remote control. The touch sensor is the start button that will begin the wrapper compression process.

Place the candy wrappers into the trash container bin sub-assembly and press the touch sensor start button. The motor starts the pneumatic compressor system that slowly lowers the pneumatic ram sub-assembly. The pneumatic ram sub-assembly exerts pressure on the trash, which flattens it. When the pneumatic ram pressure reaches its setpoint, the motor reverses and raises the pneumatic ram back up to the top of the compactor. Additional wrappers may be added and the compactor can be started again, and the process can be repeated until the bin is full. The bottom plate in the trash container bin may be removed for easy cleaning. The turntable allows the bin to be rotated as needed.

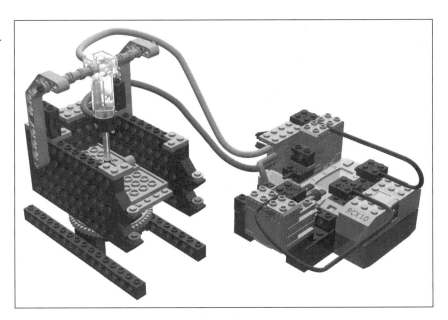

Robot 10

Robo-Hominid

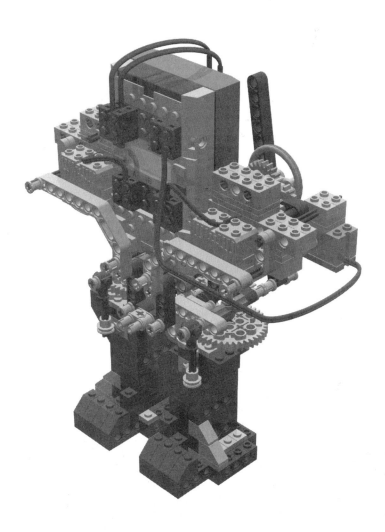

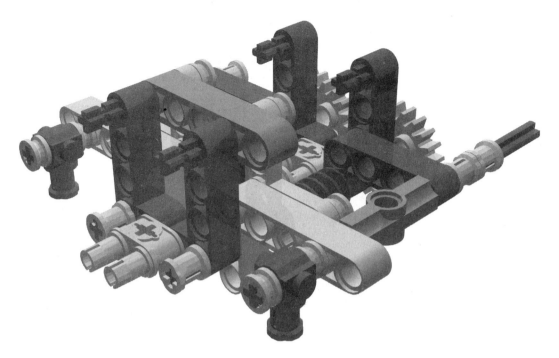

Robo-Hominid is a biped, or two legged walker. He moves by shifting his weight from side to side. Robo-Hominid is very compact for a weight shifting biped. I invented him before I was aware of the large LEGO community of adult fans. Most MINDSTORMS bipeds use a series of lift arms to build the legs; Robo-Hominid's legs are instead constructed of a number of bricks. I've not seen any designs like him in all my LEGO biped Web searching. Robo-Hominid can walk forwards, backwards and turn right and left. He has two motors, one for leaning left and right, and the other for leg striding.

Robo-Hominid uses one motor in conjunction with the rotation sensor to lean onto his left foot. He then uses the other motor to drive the right leg forward. The touch sensors on the front of the legs tell the RCX when the right leg is completely forward. The RCX then leans Robo-Hominid on his right leg, and drives the left leg forward.

You can find the program for Robo-Hominid on the Syngress Solution Web site (www.syngress.com/solutions). Robo-Hominid's program allows him to walk forward a given distance, turn himself around 180 degrees, walk back to where he started, then turn himself around again.

Bicks & Chips...

Extra Parts Needed For Robo-Hominid

Robo-Hominid uses three parts that are not found in the RIS 2.0 or the Ultimate Builders Set. The first part is a rotation sensor that comes as part of the LEGO MINDSTORMS Ultimate Accessories Kit. The other two parts are worm gears that can be found in many of the Slizer kits.

Left Leg

Begin by putting together the left leg sub-assembly.

Left Leg Step 0

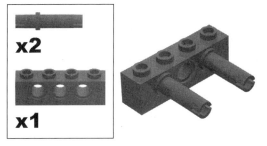

Left Leg Step 1

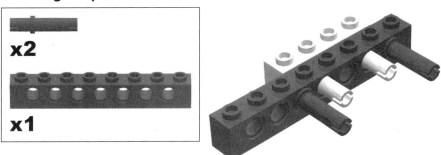

Left Leg Step 2

x2

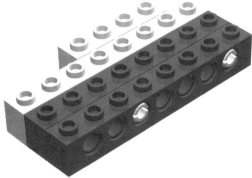

Left Leg Step 3

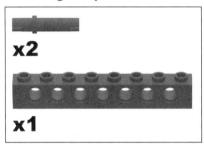

x2

x1

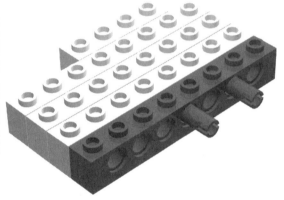

Left Leg Step 4

x1

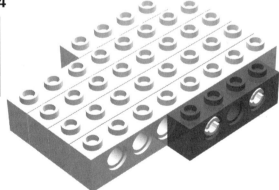

Left Leg Step 5

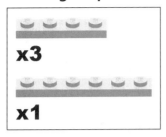

x3

x1

Left Leg Step 6

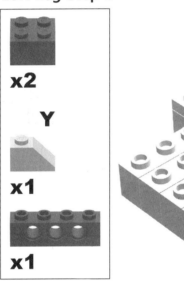

x2

Y

x1

x1

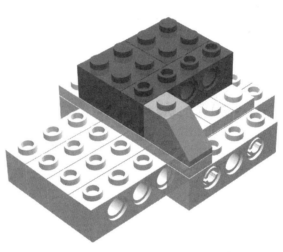

Left Leg Step 7

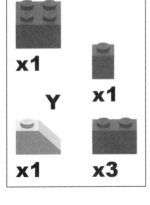

x1

Y

x1

x1

x3

Left Leg Step 8

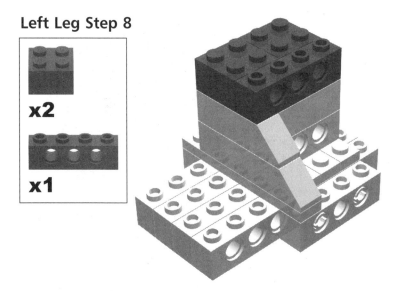

x2

x1

SYNGRESS
syngress.com

Bricks & Chips…

Movement Patterns

Try programming other patterns for Robo-Hominid to walk in, as opposed to the simple back and forth program available for him on the Syngress Solutions web site (www.syngress.com/solutions). Setting up a pre-arranged obstacle course and having Robo-Hominid try to navigate around it is a good starting point for this.

Left Leg Step 9

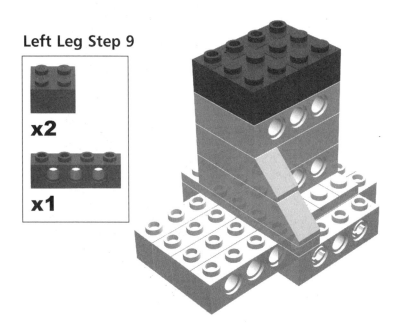

x2

x1

Left Leg Step 10

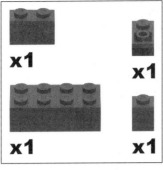

x1

x1

x1

x1

Left Leg Step 11

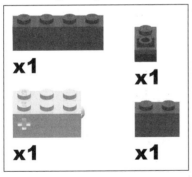

x1

x1

x1

x1

Connect an electrical cable to the touch sensor with the wire pointing to the inside of the leg. Leave the other end unconnected for now.

Left Leg Step 12

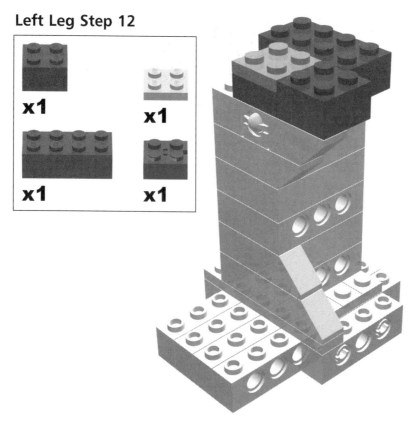

Left Leg Step 13

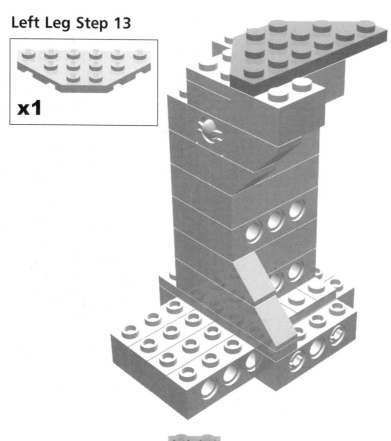

Left Leg Step 14

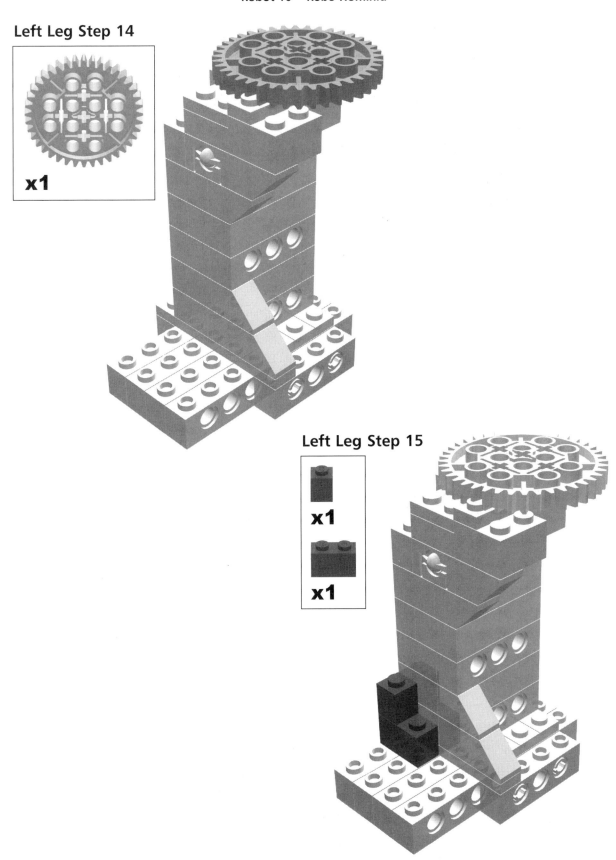

x1

Left Leg Step 15

x1

x1

Left Leg Step 16

The Right Leg

Now begin construction of the right leg sub-assembly, which is a mirror of the left leg sub-assembly.

Right Leg Step 0

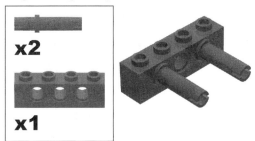

x2

x1

Right Leg Step 1

x2

x1

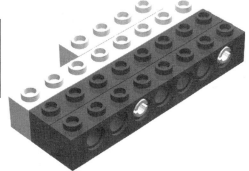

Right Leg Step 2

x2

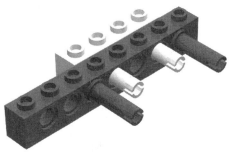

Right Leg Step 3

x2

x1

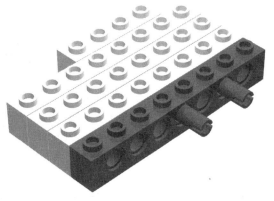

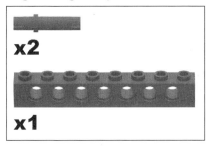

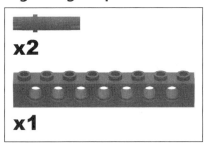

Right Leg Step 4

x1

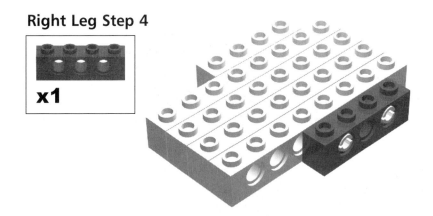

Right Leg Step 5

x3

x1

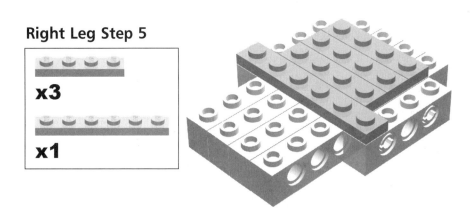

Right Leg Step 6

x2

Y

x1

x1

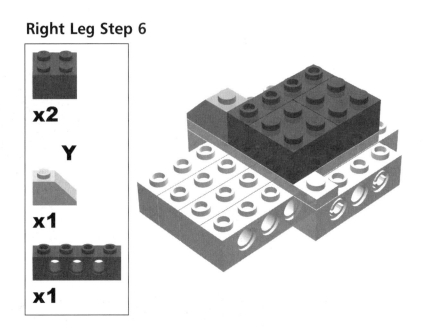

Right Leg Step 7

x1

Y **x1**

x1 **x3**

Right Leg Step 8

x2

x1

Right Leg Step 9

x2

x1

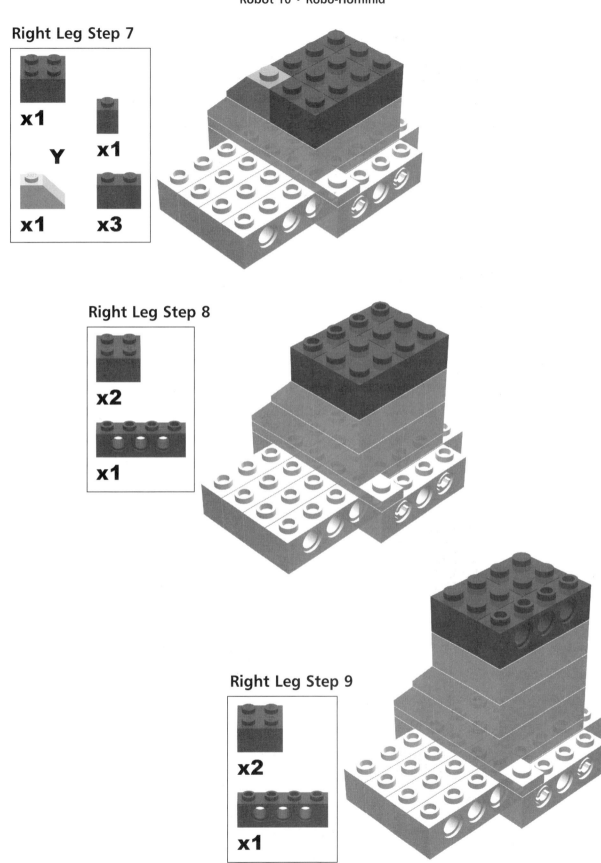

Right Leg Step 10

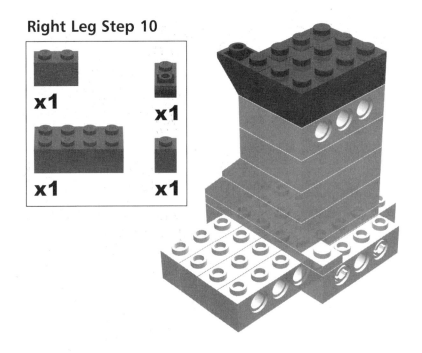

Right Leg Step 11

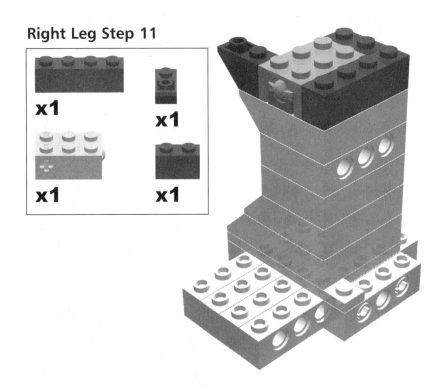

Right Leg Step 12

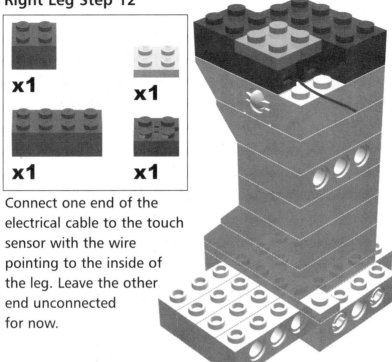

x1 **x1**

x1 **x1**

Connect one end of the electrical cable to the touch sensor with the wire pointing to the inside of the leg. Leave the other end unconnected for now.

Right Leg Step 13

x1

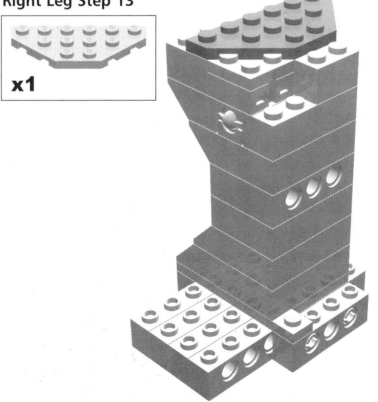

Right Leg Step 14

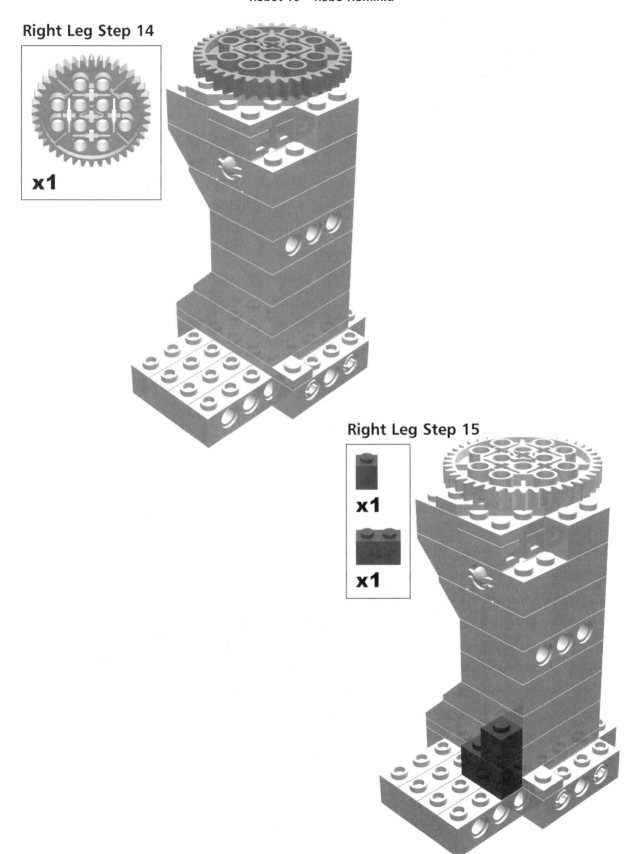

x1

Right Leg Step 15

x1

x1

Right Leg Step 16

x4

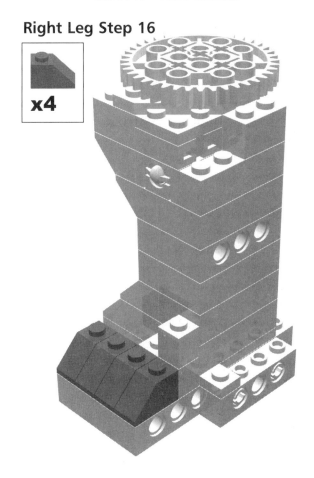

The Hips

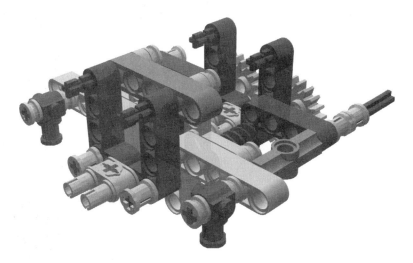

Now begin building the hips sub-assembly.

Hips Step 0

4
x1

x2

x1

x2

6
x1

x1

Y
x1

B
x1

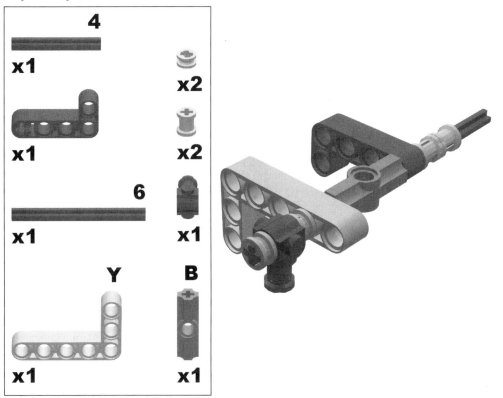

Hips Step 1

x1

x1

10
x1

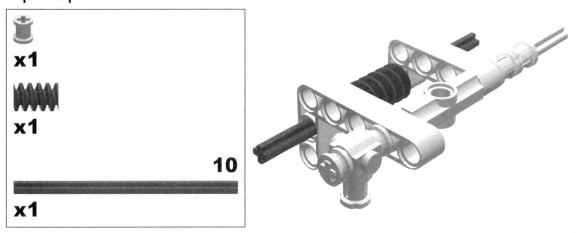

Hips Step 2

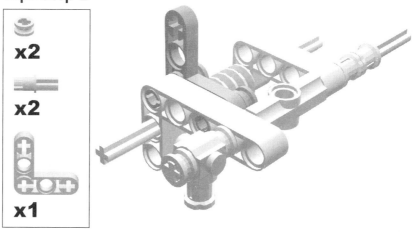

Hips Step 3

Y

2

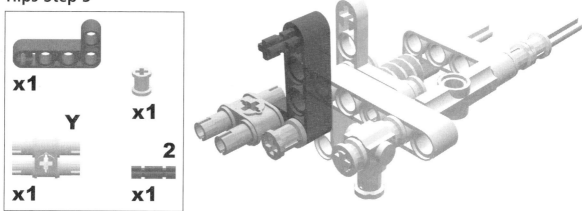

Hips Step 4

2

10

Y

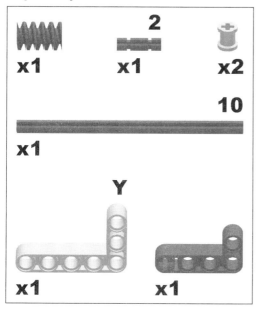

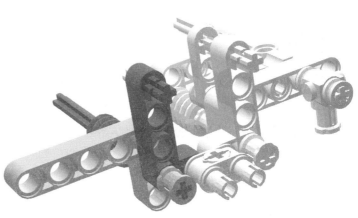

Hips Step 5

x1

3
x1

4
x1

x1

x2

x1

B
x1

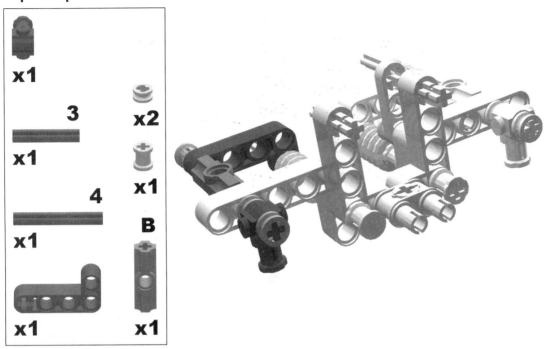

Hips Step 6

x2

x2

x1

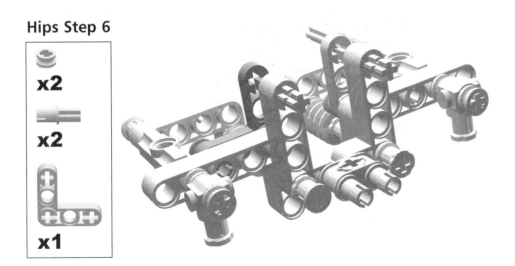

Hips Step 7

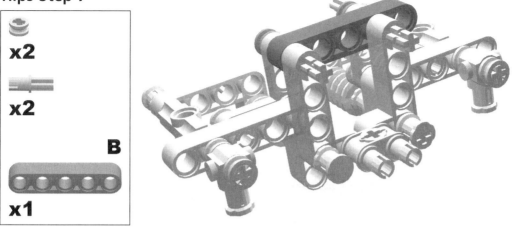

Hips Step 8

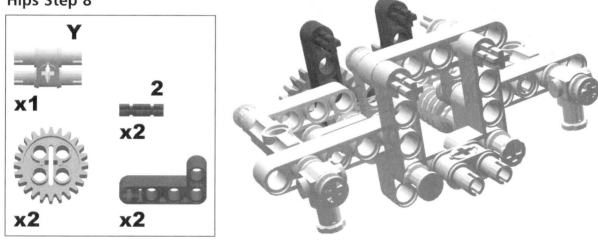

The Drive

Now begin constructing the drive sub-assembly.

Drive Step 0

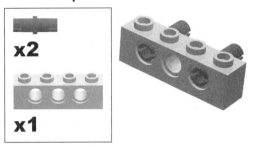

x2

x1

Developing & Deploying...
The Drive sub-assembly

The Drive sub-assembly translates power from the RCX brick to the legs.

Drive Step 1

x1

x1

x1

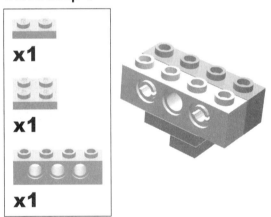

Drive Step 2

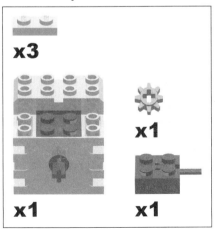

x3

x1

x1

x1

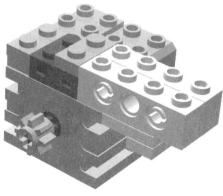

Connect an electrical cable to the top of the motor, with the wire pointing towards the back of the motor as shown. Leave the other end of the cable unconnected for now.

Drive Step 3

x1

x2

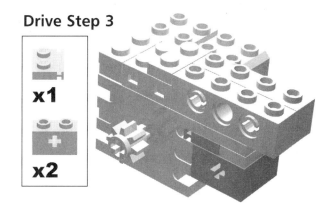

Drive Step 4

x2

x2

x1

x1

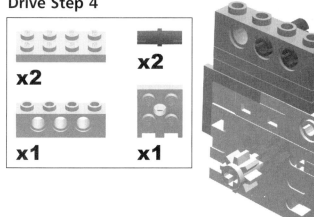

Drive Step 5

x1

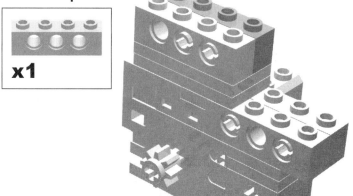

Drive Step 6

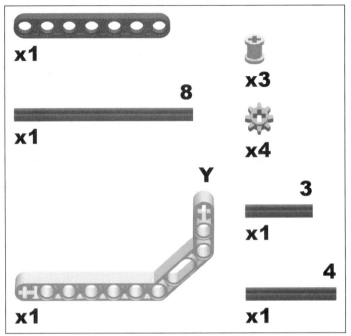

8

x1

Y

x1

x3

x4

3

x1

4

x1

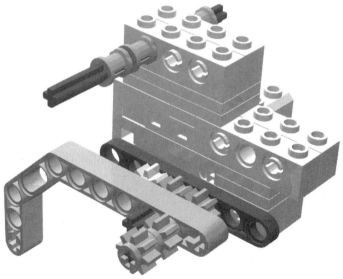

Drive Step 7

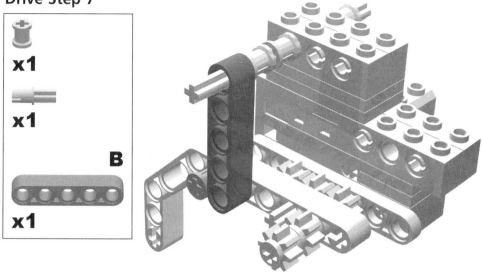

B

x1

x1

x1

Drive Step 8

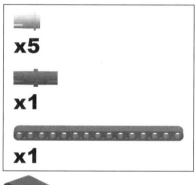

x5

x1

x1

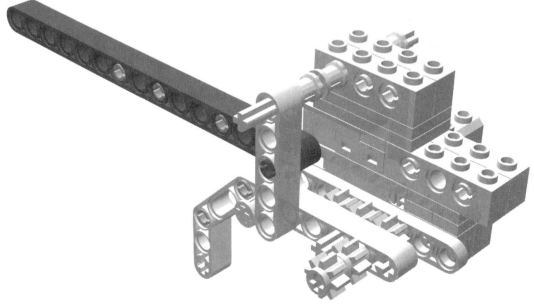

Drive Step 9

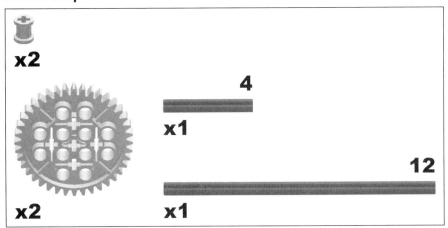

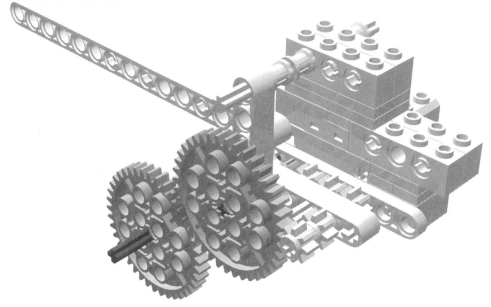

Drive Step 10

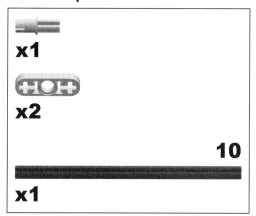

x1

x2

10

x1

Drive Step 11

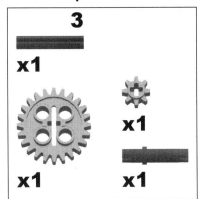

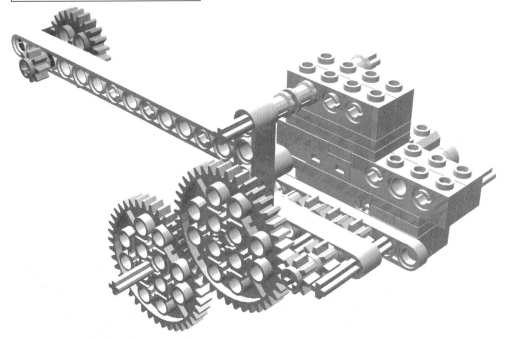

Drive Step 12

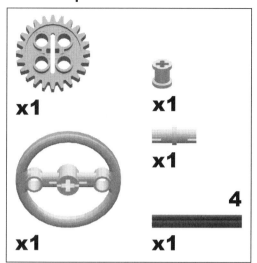

x1

x1

x1

x1

4

x1

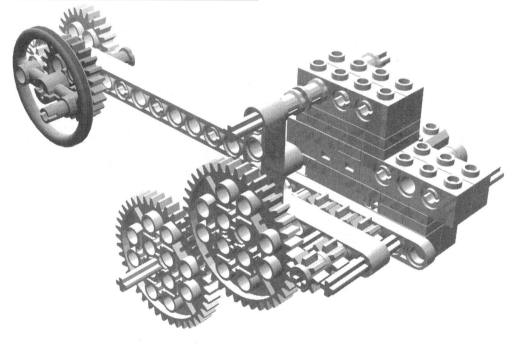

Drive Step 13

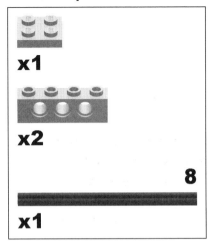

x1

x2

8

x1

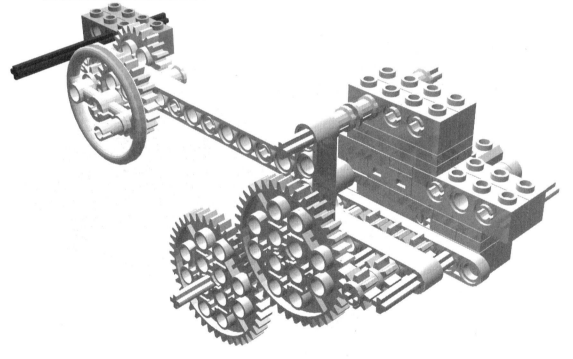

Drive Step 14

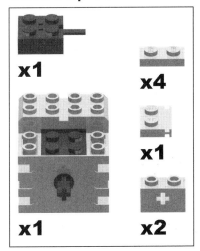

Connect an electrical cable to the top of the motor with the wire pointing towards the back of the motor. Leave the other end of the cable unconnected for now.

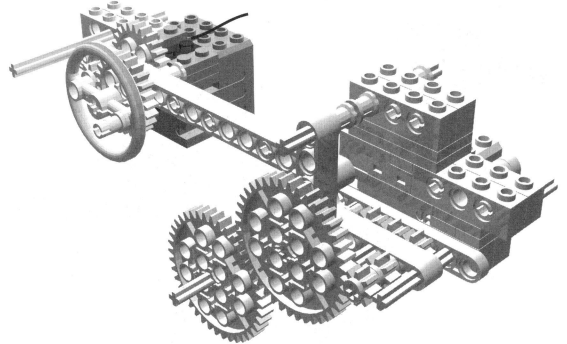

Drive Step 15

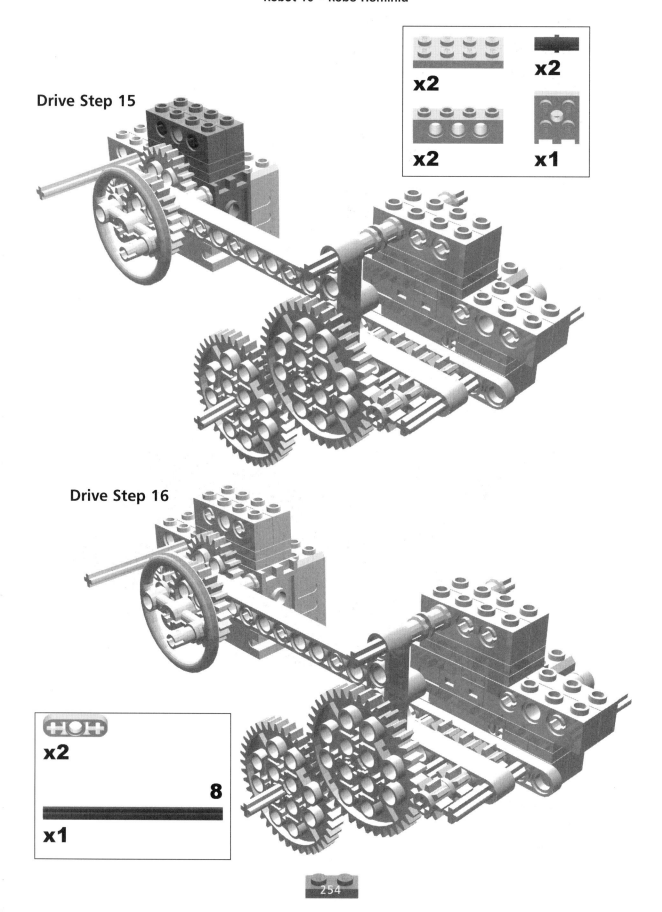

Drive Step 16

Drive Step 17

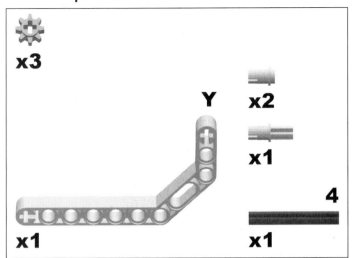

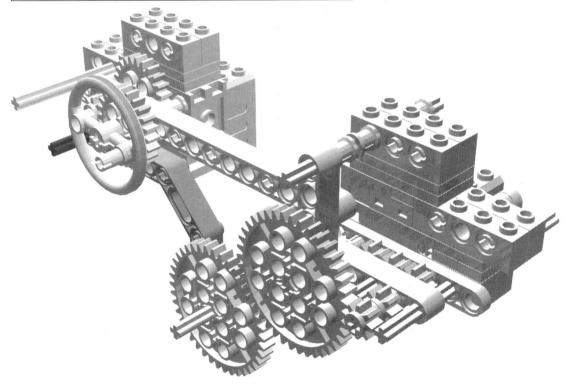

Drive Step 18

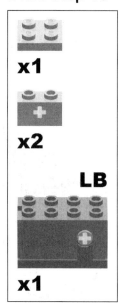

x1

x2

LB

x1

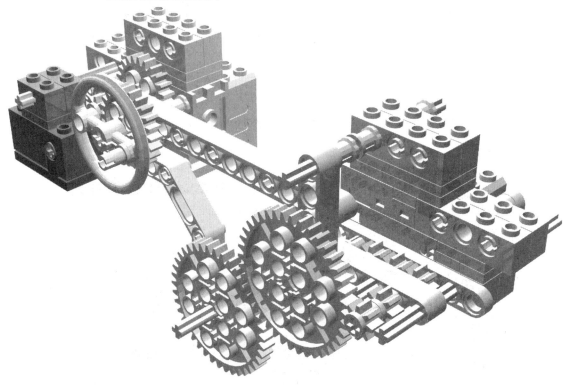

Drive Step 19

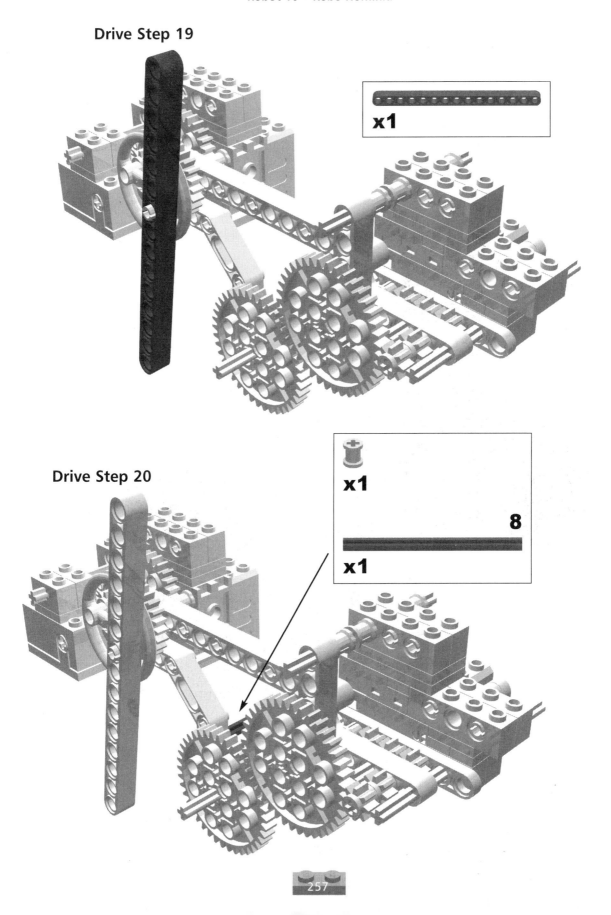

Drive Step 20

x1

8

x1

Putting It All Together

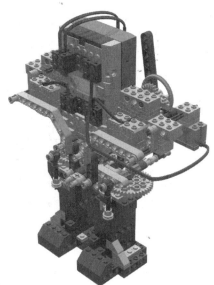

It's now time to put all of the sub-assemblies you have constructed together to complete the finished Robo-Hominid.

Final Step 0

Final Step 1

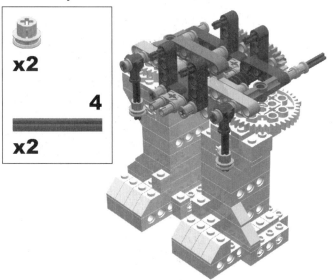

Connect the left and right leg sub-assemblies to the hips sub-assembly.

Final Step 2

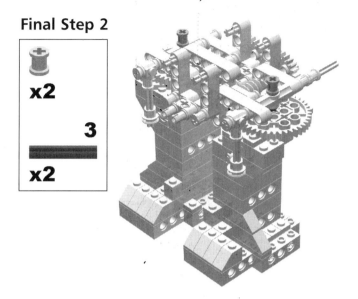

Final Step 3

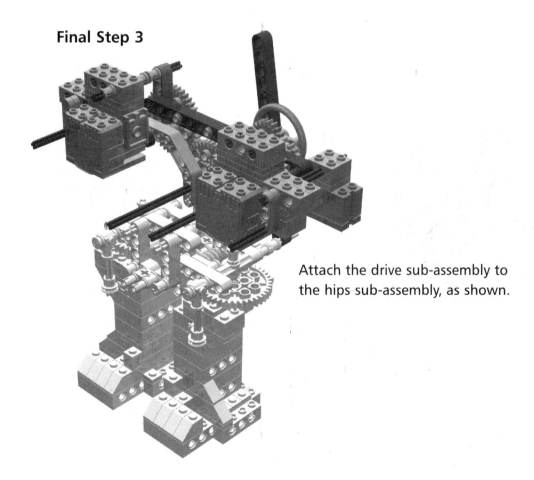

Attach the drive sub-assembly to the hips sub-assembly, as shown.

Final Step 4

Attach the RCX brick, as shown.

x4

x2

x1

Final Step 5

x3

Y

x2

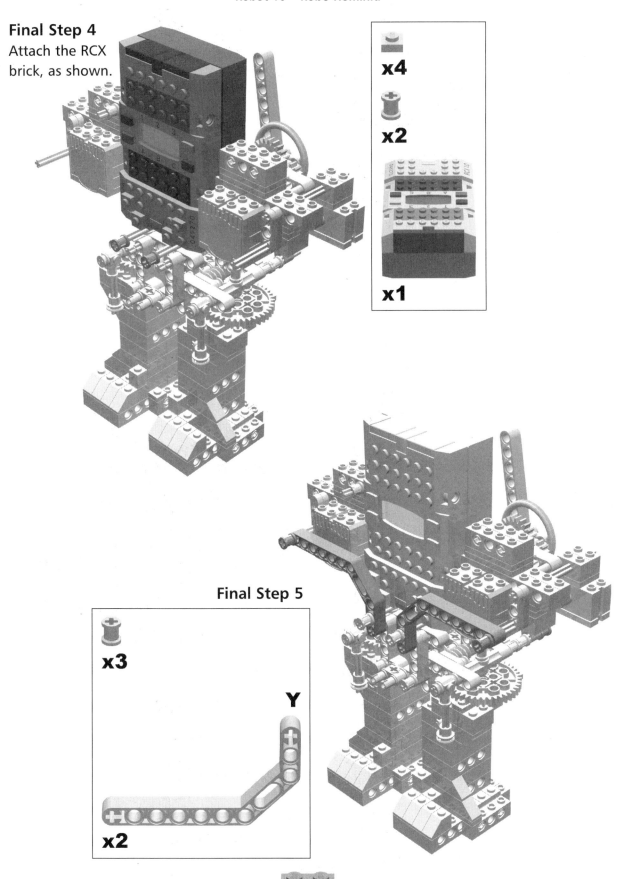

Final Step 6

x1

Hook the left leg subassembly's touch sensor cable to the Input Port 1.

Final Step 7

x1

Hook the right leg subassembly's touch sensor cable to the Input Port 1.

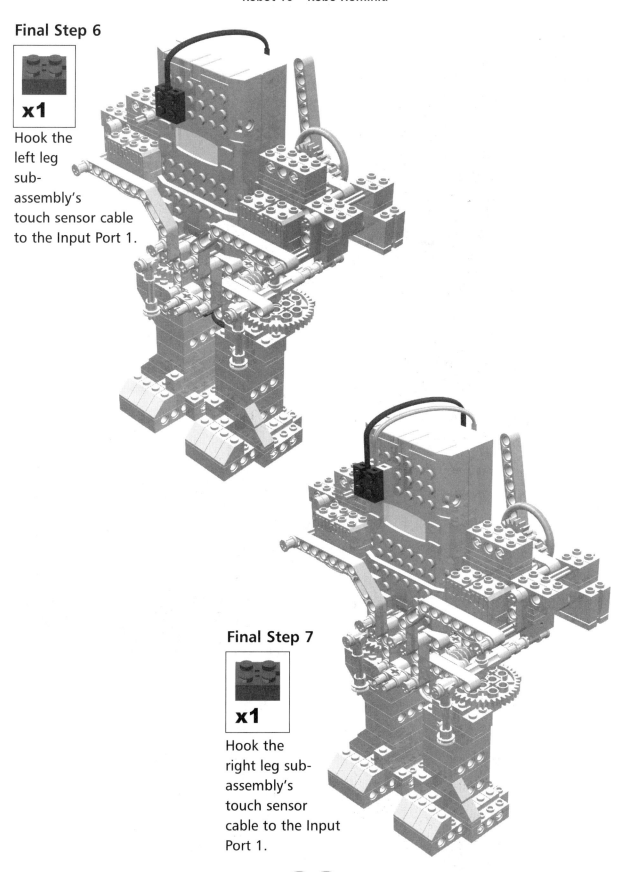

Final Step 8

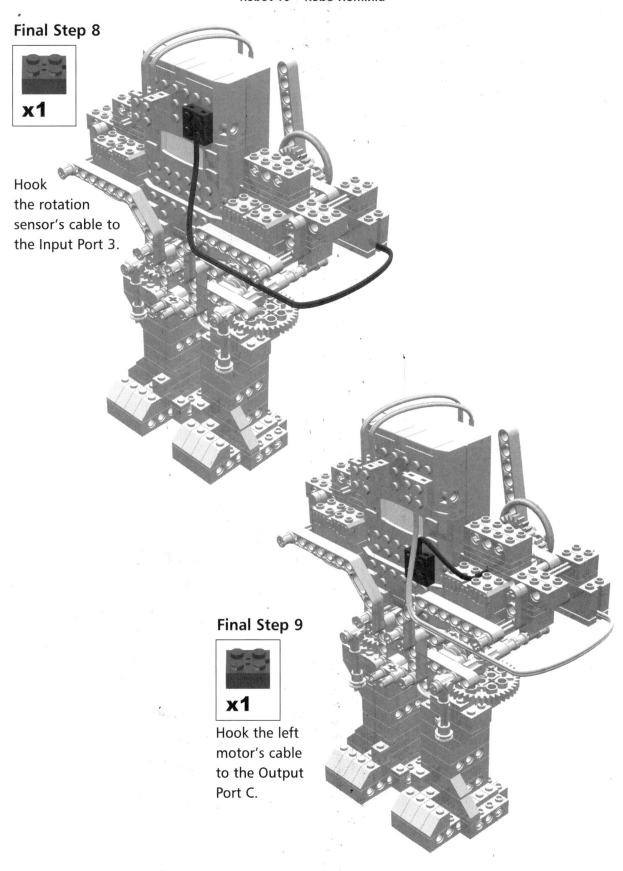

x1

Hook the rotation sensor's cable to the Input Port 3.

Final Step 9

x1

Hook the left motor's cable to the Output Port C.

Final Step 10

x1

Hook the right
motor's
cable to
the Output Port A.

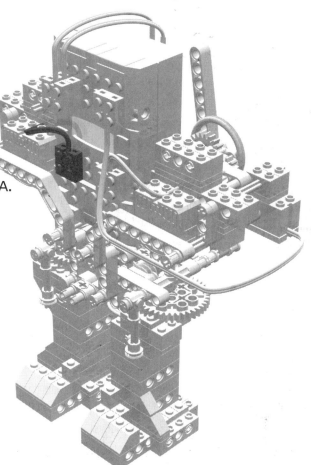

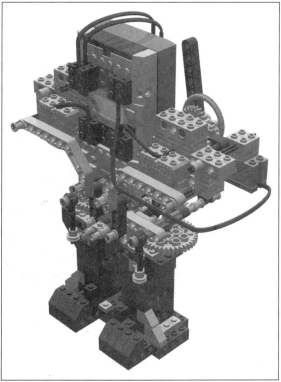

SYNGRESS SOLUTIONS...

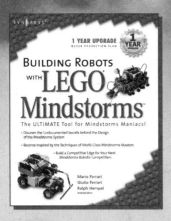

AVAILABLE NOW
ORDER at
www.syngress.com

Building Robots with LEGO MINDSTORMS

The LEGO MINDSTORMS Robotics Invention System (RIS) has been called "the most creative play system ever developed." This book unleashes the full power and potential of the tools, bricks, and components that make up LEGO MINDSTORMS. Some of the world's leading LEGO MINDSTORMS inventors share their knowledge and development secrets. You will discover an incredible range of ideas to inspire your next invention. This is the ultimate insider's look at LEGO MINDSTORMS and is the perfect book whether you build world-class competitive robots or just like to mess around for the fun of it.

ISBN: 1-928994-67-9

Price: $29.95 US, $46.95 CAN

More Great Books in the Syngress 10 Cool Series!

The 10 Cool Series covers the most popular MINDSTORMS kits from LEGO and these books give you everything you need to create cool robotics projects in under one hour.

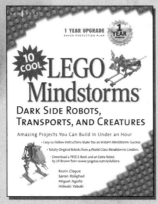

AVAILABLE NOW
ORDER at
www.syngress.com

AVAILABLE NOW
ORDER at
www.syngress.com

10 Cool LEGO MINDSTORMS Dark Side Robots, Transports, and Creatures

ISBN: 1-931836-59-0

Price: $24.95 US, $38.95 CAN

10 Cool LEGO MINDSTORMS Robotics Invention Systems 2™ Projects

ISBN: 1-931836-61-2

Price: $24.95 US, $38.95 CAN

solutions@syngress.com

SYNGRESS®